GALACTIC PIONEERS
SECURITY PASS
COMMANDER

>FIRST NAME

>LAST NAME

ACCESS ALL AREAS

VALID
FROM: NOW
TO: THE FUTURE

MISSION CONTROL

To
IMOGEN
and
LUCY

A GALAXY of HER OWN

AMAZING STORIES of WOMEN in SPACE

LIBBY JACKSON

CENTURY

CONTENTS

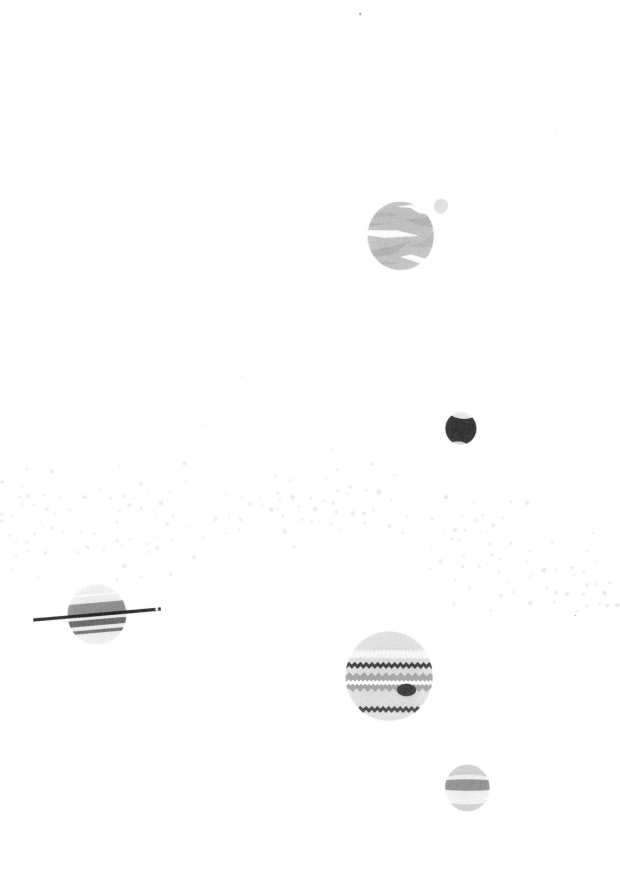

SPACE EXPLORATION TIMELINE

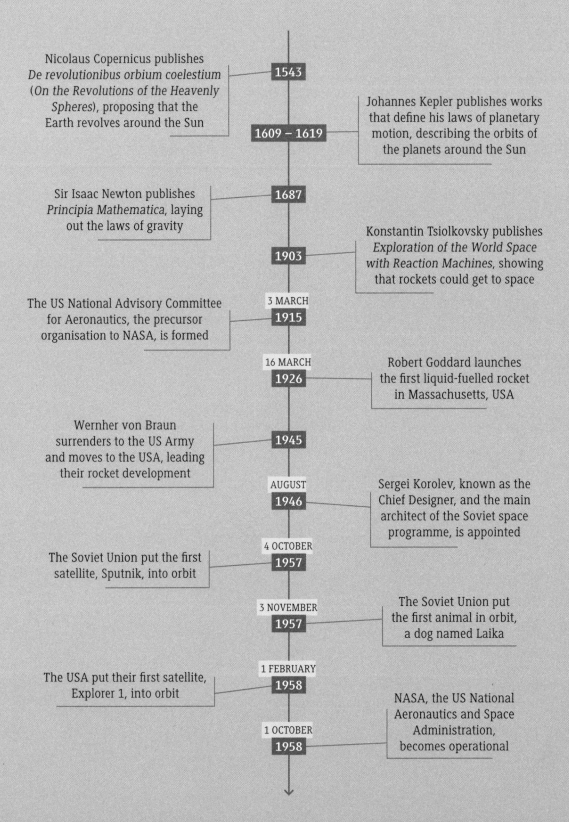

1543 — Nicolaus Copernicus publishes *De revolutionibus orbium coelestium* (*On the Revolutions of the Heavenly Spheres*), proposing that the Earth revolves around the Sun

1609 – 1619 — Johannes Kepler publishes works that define his laws of planetary motion, describing the orbits of the planets around the Sun

1687 — Sir Isaac Newton publishes *Principia Mathematica*, laying out the laws of gravity

1903 — Konstantin Tsiolkovsky publishes *Exploration of the World Space with Reaction Machines*, showing that rockets could get to space

3 MARCH 1915 — The US National Advisory Committee for Aeronautics, the precursor organisation to NASA, is formed

16 MARCH 1926 — Robert Goddard launches the first liquid-fuelled rocket in Massachusetts, USA

1945 — Wernher von Braun surrenders to the US Army and moves to the USA, leading their rocket development

AUGUST 1946 — Sergei Korolev, known as the Chief Designer, and the main architect of the Soviet space programme, is appointed

4 OCTOBER 1957 — The Soviet Union put the first satellite, Sputnik, into orbit

3 NOVEMBER 1957 — The Soviet Union put the first animal in orbit, a dog named Laika

1 FEBRUARY 1958 — The USA put their first satellite, Explorer 1, into orbit

1 OCTOBER 1958 — NASA, the US National Aeronautics and Space Administration, becomes operational

INTRODUCTION

I have been fascinated by space my whole life. Like many of the women in this book, I remember looking up in awe at the night sky as a child, learning the constellations, hoping to catch sight of a shooting star. I was captivated by the Moon, bright and beautiful, by the stories of those who had walked on it years before I was even born, and by the Space Shuttles that were roaring into orbit.

But I didn't long to be an astronaut or to work in the space industry – to me, those were jobs done in America, a world away from the suburbs of London where I was growing up. I simply enjoyed solving puzzles, trying to understand what I observed around me, and playing with machines and computers.

Three decades later, and I've taken my fascination with space and turned it into a career that has fulfilled and surpassed anything I ever thought possible back then. The story of how I went from one to the other is part ambition, part determination and part good fortune, but it is a path that I believe anyone can tread, just as the amazing women in this book have done.

At school, I liked maths, science, music, and learning how the world worked. Aged ten, I followed eagerly when the papers were full of the first British astronaut going to space – Helen Sharman. Six years later, my physics teacher, Mr Farrow, held up a yellow leaflet and asked if anyone was interested in going to something that sounded very exciting, Space School. Generously my parents agreed that I could go. So I trundled off on my own that summer, and spent a week enthralled by lessons in rocketry and engineering. To my amazement I even visited a company in the UK which made satellites. Slowly, I started to see that my passion for space might actually lead to a job, something that had never crossed my mind before.

In the first year of A levels, my friends and I were tasked with finding placements for work shadowing. I clearly remember us sitting in the common room, discussing what we would like to do. Some were writing to doctors, lawyers and vets, others to musicians or theatres.

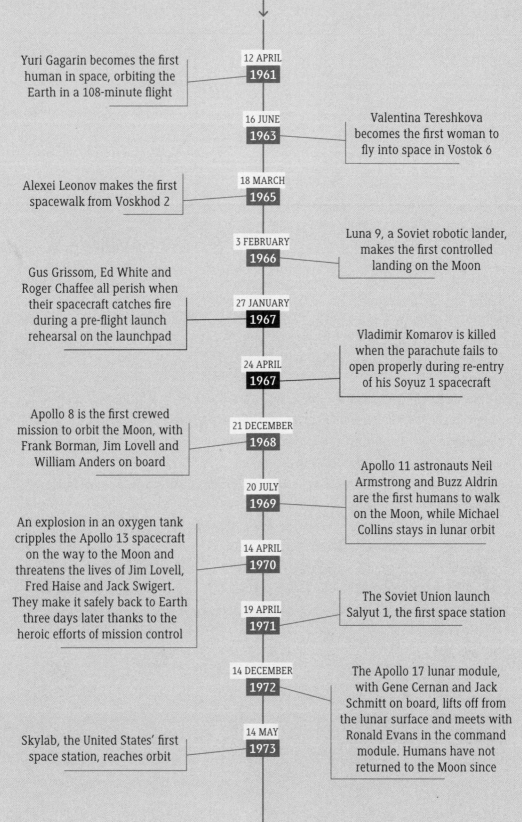

Yuri Gagarin becomes the first human in space, orbiting the Earth in a 108-minute flight

12 APRIL 1961

16 JUNE 1963

Valentina Tereshkova becomes the first woman to fly into space in Vostok 6

Alexei Leonov makes the first spacewalk from Voskhod 2

18 MARCH 1965

3 FEBRUARY 1966

Luna 9, a Soviet robotic lander, makes the first controlled landing on the Moon

Gus Grissom, Ed White and Roger Chaffee all perish when their spacecraft catches fire during a pre-flight launch rehearsal on the launchpad

27 JANUARY 1967

24 APRIL 1967

Vladimir Komarov is killed when the parachute fails to open properly during re-entry of his Soyuz 1 spacecraft

Apollo 8 is the first crewed mission to orbit the Moon, with Frank Borman, Jim Lovell and William Anders on board

21 DECEMBER 1968

20 JULY 1969

Apollo 11 astronauts Neil Armstrong and Buzz Aldrin are the first humans to walk on the Moon, while Michael Collins stays in lunar orbit

An explosion in an oxygen tank cripples the Apollo 13 spacecraft on the way to the Moon and threatens the lives of Jim Lovell, Fred Haise and Jack Swigert. They make it safely back to Earth three days later thanks to the heroic efforts of mission control

14 APRIL 1970

19 APRIL 1971

The Soviet Union launch Salyut 1, the first space station

14 DECEMBER 1972

The Apollo 17 lunar module, with Gene Cernan and Jack Schmitt on board, lifts off from the lunar surface and meets with Ronald Evans in the command module. Humans have not returned to the Moon since

Skylab, the United States' first space station, reaches orbit

14 MAY 1973

Someone asked what I would like to do and without blinking I replied, 'I want to work at NASA one day.' Two of us hatched a plan – we would email NASA – not for a moment thinking it would succeed. A couple of weeks later, we were flabbergasted. Not only had we got a reply, but they had said yes! So in March 1998 we set off for two weeks in Houston.

Out of this world doesn't even begin to cover the visit, and all the brilliant things we saw, from the Moon rocks to spacewalk training and many things in between. But I will never, ever, forget walking into mission control, sitting next to Cathy Larson, the propulsion engineer, and watching the team practise Shuttle launch-and-abort scenarios. As soon as I put on the headset, heard the flight director bring the team of hugely talented people together, and watched as they harmoniously responded to problems in real time, I knew this was where I belonged.

I came back from that trip with a new dream – to work in mission control. How that was going to happen, I had no idea. I was a Brit and NASA only hired US citizens. The British government did not support human space flight, so after Helen Sharman surely we'd never have anything as exciting as another astronaut. I quietly filed my ambitions away, thinking it was most likely impossible.

I went off to study physics at university, then a fascinating master's degree in space engineering, before joining a graduate training scheme at Astrium, the satellite company I had visited years before. I spent three years installing a new control centre for some communications satellites, and preparing for their launch, loving every moment of my work. But as I learnt that the European Space Agency (ESA) was going to become part of the International Space Station (ISS) and was preparing to launch a new scientific laboratory into orbit called Columbus, I became restless and started thinking about how I could be a part of the programme.

When I saw an advert for an instructor at the Columbus Control Centre in Germany, I applied and was very excited when I was asked to visit for an interview. After being quizzed on my abilities, I was given a tour of the facility. As my guide held his pass up to enter the control room, my heart raced. When I stepped into the cavernous room, I had the same feeling I had had a decade earlier in Houston – the peace and tranquillity of a control room, bathed in the glow of dozens of computer monitors, the beating heart of a space mission. Though I spoke no German and had no clue what moving to Europe would involve, when they offered me the job I knew I had to take it.

As soon as I began training flight controllers, I told anyone who would listen that I wanted to be one myself, and that one day I wanted to be a flight director. My managers took note and my tenacity paid off – before long, I was going through the months of intense training. I finally took my seat in mission control as a data and communications flight controller. Then three years later my impossible dream became a reality – I was in charge of a control room, overseeing day-to-day operations as a Columbus flight director.

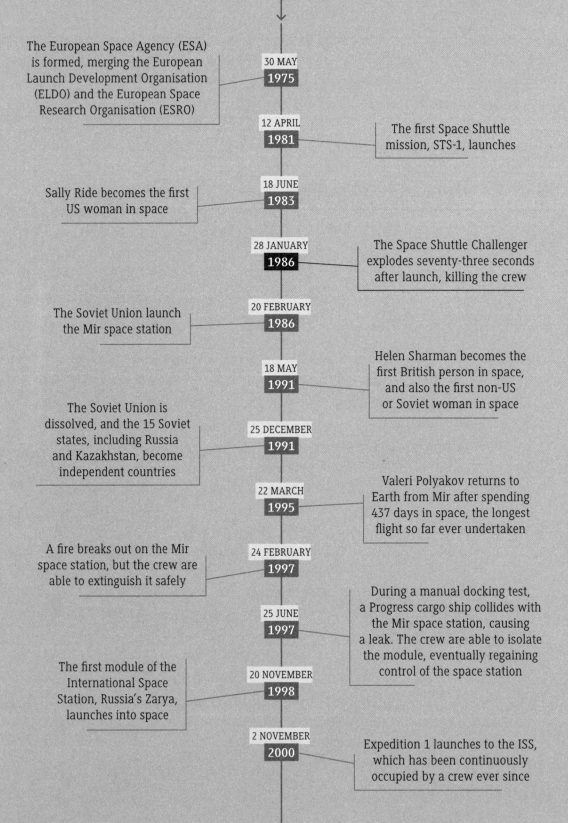

30 MAY 1975
The European Space Agency (ESA) is formed, merging the European Launch Development Organisation (ELDO) and the European Space Research Organisation (ESRO)

12 APRIL 1981
The first Space Shuttle mission, STS-1, launches

18 JUNE 1983
Sally Ride becomes the first US woman in space

28 JANUARY 1986
The Space Shuttle Challenger explodes seventy-three seconds after launch, killing the crew

20 FEBRUARY 1986
The Soviet Union launch the Mir space station

18 MAY 1991
Helen Sharman becomes the first British person in space, and also the first non-US or Soviet woman in space

25 DECEMBER 1991
The Soviet Union is dissolved, and the 15 Soviet states, including Russia and Kazakhstan, become independent countries

22 MARCH 1995
Valeri Polyakov returns to Earth from Mir after spending 437 days in space, the longest flight so far ever undertaken

24 FEBRUARY 1997
A fire breaks out on the Mir space station, but the crew are able to extinguish it safely

25 JUNE 1997
During a manual docking test, a Progress cargo ship collides with the Mir space station, causing a leak. The crew are able to isolate the module, eventually regaining control of the space station

20 NOVEMBER 1998
The first module of the International Space Station, Russia's Zarya, launches into space

2 NOVEMBER 2000
Expedition 1 launches to the ISS, which has been continuously occupied by a crew ever since

My job brought together two of my great passions: solving puzzles and space. I led the mission control team during shifts, keeping the crew and the spacecraft safe whilst we worked to help them accomplish their daily planned activities. I relished simulations, when we were challenged with problem after problem. Most days in the simulated version of the space station ended with experiments that refused to work properly, malfunctioning life support and data systems, and often a fire or a water leak. The practices made sure we could handle real problems confidently and safely, and help the crew get the science experiments done every day.

I was just thirty years old, with the job of my dreams. What on earth would I do next?

In 2013, on holiday with a friend, I was waiting in the baggage reclaim hall at Munich airport and idly scrolling through Twitter. Suddenly I squealed. My friend looked over quizzically. 'Tim Peake is going to space!' I all but shouted across the airport. In November 2012, the UK government had somewhat unexpectedly decided to contribute to the ISS. Just six months later, much sooner than I had thought would happen, Tim Peake had been assigned a flight – finally a second astronaut representing the UK was going into space. I simply had to be part of the mission.

When the UK Space Agency advertised for someone to manage their education and outreach programme for Tim's flight, I knew, just knew, that the job was perfect for me. Throughout my career, I had always believed that the inspirational value that space, and particularly astronauts, gives young children is priceless. Although I didn't have all the qualifications listed, I applied, hopeful that if I could just get an interview I could show them what they were missing.

One dreary December day I made my way to the UK Space Agency in Swindon. I had done my homework and came with a vision for a space education programme like nothing before. A few weeks later, my preparation and passion paid off and to my utter elation, I was offered the job. The Union Jack was on an astronaut's arm once more, and supported by colleagues from the UK Space Agency and a host of other organisations, we planned and delivered a hugely successful educational programme – determined that children from all around the country would have the opportunity to learn and be inspired by science and space.

I now manage human space flight and microgravity science in the UK, something that throughout my life seemed unimaginable for anyone to do, let alone me. Today though, space isn't just for those in the USA or Russia; it is a worldwide industry that underpins everyday life, from weather forecasting to satellite navigation, communications to exploring the solar system. The sector is growing and needs young people to follow their dreams and join it.

I've never lost sight of my own dreams, have seized every opportunity, and always worked as hard as possible to do my very best. I'm so proud of all I've done in my career, but some of my most treasured achievements are from life outside of work. I was terrible at

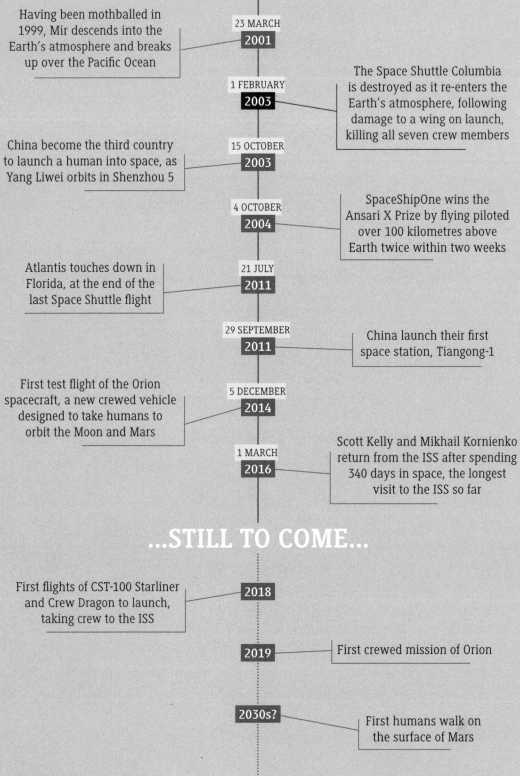

Having been mothballed in 1999, Mir descends into the Earth's atmosphere and breaks up over the Pacific Ocean

23 MARCH 2001

The Space Shuttle Columbia is destroyed as it re-enters the Earth's atmosphere, following damage to a wing on launch, killing all seven crew members

1 FEBRUARY 2003

China become the third country to launch a human into space, as Yang Liwei orbits in Shenzhou 5

15 OCTOBER 2003

SpaceShipOne wins the Ansari X Prize by flying piloted over 100 kilometres above Earth twice within two weeks

4 OCTOBER 2004

Atlantis touches down in Florida, at the end of the last Space Shuttle flight

21 JULY 2011

China launch their first space station, Tiangong-1

29 SEPTEMBER 2011

First test flight of the Orion spacecraft, a new crewed vehicle designed to take humans to orbit the Moon and Mars

5 DECEMBER 2014

Scott Kelly and Mikhail Kornienko return from the ISS after spending 340 days in space, the longest visit to the ISS so far

1 MARCH 2016

...STILL TO COME...

First flights of CST-100 Starliner and Crew Dragon to launch, taking crew to the ISS

2018

First crewed mission of Orion

2019

First humans walk on the surface of Mars

2030s?

languages, my English teachers despaired of me, and I was always the last to be picked for any sports team. Since then I've moved to another country and learnt to speak the language, I've run the London Marathon – twice – and now, to my amazement, I've written a book. With hard work and determination anyone can achieve anything they set their mind to.

This book, so beautifully illustrated by students from the London College of Communication, tells the extraordinary stories of over fifty talented and heroic women from throughout history who have all played their part in humankind's journey into the stars above. In a field that has historically been dominated by men, it is vitally important to celebrate the achievements and contributions of women to remind us all, particularly young people, that anyone can follow their dreams into the world of space. The historical dominance of the USA and the Soviet Union in the early days of exploration mean that many of the women featured are from these countries, but this is not a reflection of the industry today. Space is by its very definition an international arena and in such a large undertaking, countries find that working together can be much more beneficial to all.

There are, of course, thousands of women worldwide who play and have played vital roles in the story of human space flight, and to pick out just fifty in this book was incredibly difficult. Many more engineers, scientists, doctors, lawyers, managers, technicians, astronauts and people in countless other jobs have made superb contributions, and the fact that they are not included here is no reflection on their abilities or achievements.

There are some common themes that run through the stories in this book. The women, regardless of the period of history they were living in, refused to be limited by any barriers that society tried to place upon them, and wouldn't listen to anyone suggesting that they couldn't or shouldn't strive to fulfil their ambitions. They have followed their passions, grabbed opportunities whenever they arose, stayed motivated and always endeavoured to do their very best. It wasn't possible to include all of the amazing details of their lives in these pages, but I hope I am able to give you a taste of their remarkable achievements. I urge you to find out more about them, and indeed the many other people who have made human space flight possible. I promise you will be enthralled by their stories.

Remember, find your passion in life, whatever it might be. Never lose sight of it and don't be afraid to tell anyone what it is. Seek out opportunities and grab them wholeheartedly. Don't shy away from making difficult decisions, and never be scared to ask. You can do anything you set your mind to. Above all, have fun and enjoy all you do, life is too short for anything else.

When I left school, my head teacher wrote in everyone's yearbook, 'The sky's the limit!', but then she said, 'I don't think that's true'. It's the same for everyone. The whole universe is out there and it's waiting for you.

LIBBY JACKSON, August 2017

THE
ORIGINS
OF
SPACE
TRAVEL

THE ORIGINS OF SPACE TRAVEL

$\longrightarrow$ 1957

Humans have always looked up at night and wondered what all the glimmering dots of light were and how they moved across the sky. As knowledge and science progressed, Nicolaus Copernicus realised in the 1500s that the Earth was not the centre of the universe. A century or so later, Johannes Kepler and Sir Isaac Newton worked out how and why the planets move around the Sun and started a scientific revolution. Technology developed to enable us to take to the sky in flying machines, and as is human nature, people wanted to fly higher and faster, setting their sights on reaching the stars.

ÉMILIE du CHÂTELET

$\sqrt[x]{\pi}$ MATHEMATICIAN   ⚛ PHYSICIST

🌓 FRANCE   1706 → 1749

Émilie du Châtelet was born in Paris in an age when women were not allowed to visit public libraries, let alone think of studying subjects like science or go to university. But Émilie didn't let this get in her way. Through sheer courage and willpower, she went on to share groundbreaking scientific theories with the world, theories that still help us to understand space today.

Émilie was married when she was nineteen to a grand marquis chosen by her parents. As a wife she was expected to stay at home, have children and tend to the household, but she made sure she also continued to study her great passions of maths and physics, and her curiosity to learn soon had her escaping into town to explore the world about her. Once she even dressed as a man so that she could enter a café to meet with other scientists and mathematicians.

Émilie met the great philosopher and historian Voltaire, who introduced her to the books of great British scientist Sir Isaac Newton. Newton had made numerous significant discoveries, and in his most important book, *Principia Mathematica*, laid down the fundamental laws of physics. He explained the world around us, including how gravity works, and how objects in space orbit the Earth. Newton's theories remained controversial at the time, and were extremely complex, but Émilie refused to be daunted, and spent years studying them.

She was in the middle of writing a translation of *Principia Mathematica* so that the whole of France – and not just the educated men – could understand these radical ideas, when she discovered she was pregnant. Émilie was forty-two and, despite having given birth three times already, feared she might not survive childbirth at such an age. But as always, she showed a powerful determination, working all the hours she could to complete the book just a few days before her daughter Stanislas-Adélaïde was born. Émilie survived the birth, but tragically died six days later. Her great work wasn't published for another decade, but when it was, it contributed to a scientific revolution across Europe, and to this day remains the most widely read version of Newton's book in French.

TRANSLATING NEWTON'S GENIUS

Émilie's heroic efforts laid the foundations of science and space for generations. But she also showed women of her own day what was possible. Knowledge wasn't just for men – in her career she read, wrote books and gained international fame for her mighty mind.

'Let us be certain of what we want
to be; let us choose for ourselves
our own path in life.'

ADA LOVELACE

$\sqrt[x]{\pi}$ MATHEMATICIAN

✚ UNITED KINGDOM   1815 ⟶ 1852

Ada Lovelace adored machines. When she was twelve years old, she sketched out ideas for a steam-powered flying machine that would soar through the air. Her colourful imagination and fascination with science would go on to make this young dreamer famous throughout the world.

Ada was fortunate. Although she was born at a time when it was very unusual for girls to study maths, her mother had been taught maths and insisted that her daughter was too. Ada was extremely talented and enjoyed studying languages as well as science, which would prove to be very useful.

When she was seventeen, she met Charles Babbage, a pioneer of 'thinking machines'. Babbage had invented an amazing machine, which he called the Difference Engine and was designed to fill an entire room. Today, we would think of it as a huge mechanical calculator. He then had an idea for an even more complex machine, called the Analytical Engine, which could perform more complicated calculations. Ada was captivated by Babbage's work and when an

THE FIRST PUBLISHED COMPUTER PROGRAMMER

Italian engineer wrote a paper about this new machine, she translated it into English. She was so fascinated by it that she added her own thoughts as well. In fact, she had so many ideas that the English version ended up being nearly three times as long as the original.

Ada published her work in 1843 – under the initials A.A.L. so that people would not know she was a woman and therefore think less of it – and so became the first published computer programmer. Her fizzing imagination saw the incredible potential of the machine, suggesting that it could manipulate symbols and music. Ada saw that it could also do many of the things a modern computer can and she was the first person to write out complete instructions for it. However, Babbage's great machine was never built and Ada's work was soon forgotten. It wasn't until the 1950s that the first person to build a programmable computer, Alan Turing, came across it and recognised its brilliance.

As the computing age dawned, people finally saw what a great and pioneering mind Ada had possessed, and in the 1980s a new computer language, Ada, was named after her in recognition of her work. This is still in use today, not least on some of the International Space Station's many computers.

Ada's sheer brilliance is now celebrated every year, on Ada Lovelace Day, when people all around the world mark the achievements of Ada and all of the women in science, technology, engineering and maths (STEM). She will never be forgotten again.

'Your best and wisest refuge from
all troubles is in your science.'

JEANNETTE PICCARD

HIGH-ALTITUDE BALLOONIST   SCIENTIST

USA   1895 → 1981

In the early days of flight, as the world wondered what might lie beyond the clouds, explorers flew as high as they could in balloons that were lighter than air. Jeannette Piccard and her husband Jean were passionate about science and learning. Jean wanted to study the cosmic rays in the higher levels of the atmosphere, and in order to reach these dizzying heights, he needed to be flown in a stratospheric balloon. Jeannette rose to the challenge and became the first woman to get a balloon licence, flying with great skill so that Jean could carry out important scientific observations.

Their record-breaking flight took place on 23 October 1934. More than 600 people helped to release the ropes holding down their enormous hydrogen-filled balloon, and 45,000 people

THE FIRST WOMAN IN THE STRATOSPHERE

watched them soar off into the atmosphere. Along with the Geiger counters for detecting radiation and other scientific equipment, the Piccards took their pet turtle, Fleur de Lys, for the ride. Jeannette was completely in control of flying the balloon and piloted it for eight hours, travelling almost 500 kilometres. They reached an incredible altitude of 17,550 metres above the Earth's surface – higher than any woman had travelled before.

Some people have called Jeannette the first woman in space as she travelled so high, but the definition of where space begins is at an imaginary point called the Karman line, 100 kilometres above the surface of the Earth. At about this height, it becomes impossible for aeroplanes to fly because the atmosphere is too thin. However, the Karman line is not where Earth's atmosphere stops and space begins – it actually carries on getting thinner and thinner the further up you go. The top layer of the atmosphere, called the exosphere, extends to about 10,000 kilometres above the surface of the Earth.

Jeannette and Jean's flight did not go entirely to plan. Jeannette had to change the flight path due to bad weather conditions, which meant that they did not gather all the scientific data they had hoped for. She also had to land in trees, tearing the balloon beyond repair and causing the gondola to fall the last few metres. She later said, 'What a mess! I wanted to land on the White House lawn.' Nevertheless, Jeannette's achievements were still monumental and her record for flying higher than any woman had before would not be broken for nearly thirty years, when humans finally made it into space.

'Like the hundreds of thousands of men and women ... working in the space programme, one gets a very rewarding feeling to realise that you have helped give history a little nudge forward towards the wonders of the future.'

MARY SHERMAN MORGAN

ROCKET SCIENTIST   CHEMIST

USA 1921 → 2004

Mary Sherman Morgan was a rocket scientist who worked on a space programme so secret that information about her has been in danger of being lost to time.

Mary grew up on a farm in North Dakota, and didn't start school until she was eight as her parents were reluctant to lose her valuable help with the land and animals. But she loved learning so much she ran away from home to attend university.

At university during the Second World War, her brilliance at chemistry meant she was offered a job with top-secret clearance. Although it meant she had to leave university before she had finished her studies, Mary decided to do her part for the war effort and accepted, not knowing what the job entailed. On her first day, she discovered it was at one of the world's largest makers of explosives for the US military.

At the end of the war, the USA and the Soviet Union both started using the knowledge they had learnt from weapons to develop space rockets. The Soviet Union took the first big prize when they launched the first-ever satellite, called Sputnik, which was not much bigger than a beachball, but the USA now found themselves behind in the race. The American team, led by Wernher von Braun, developed a rocket to launch their own satellite, but it wasn't powerful enough to make it all the way into space. The only way to solve the problem was to fill it with better rocket fuel, but none existed.

HER ROCKET FUEL LAUNCHED AMERICA'S FIRST SATELLITE

Mary's skills had been snapped up by a company called North American Aviation (NAA). Out of its 900 engineers she was the only woman, but she did not let that faze her and soon became the company star, designing new fuels and explosives. When the American space team asked NAA for their best man to develop a powerful rocket fuel, they were told it was Mary. They were incredulous – somebody with no university degree, and worse still, a woman – how could she possibly be up to the job? But Mary's bosses insisted she was the best.

Sure enough, she soon had the answer: a new mix of propellant she called Hydyne. The fuel was carefully transported to Cape Canaveral, Florida, and loaded into the rocket which on 31 January 1958 took Explorer 1 into space. The USA were finally in the space race. Mary, America's first female rocket scientist, had saved their early space programme.

Mary retired a few years later, and was given a hero's send-off. As she looked around the room, she was proud to see other female engineers in the crowd, following in her footsteps.

'Please, Daddy – I want to go to school.'

JACQUELINE COCHRAN

🛬 PILOT 💡 ENTREPRENEUR

🇺🇸 USA 1906 → 1980

Jaaqueline Cochran loved clothes and make-up, and dreamt of a glamorous lifestyle. She moved to New York to make her fortune, where her colourful and commanding personality soon saw her as stylist in a hairdressing salon in Saks Fifth Avenue. She made many friends and married successful businessman Floyd Odlum, who suggested that Jackie should learn to fly. As soon as Jackie was in the air, she knew it was her home and wondered why she had waited so long. She was a natural – just three weeks later she had her pilot's licence and was soon entering competitions.

Jackie was the first woman to enter the famous Bendix race across America, and in 1938 she won it. Early female pilots were constantly told they could not handle fast planes but Jackie was undeterred and set records for flying higher, faster and further than anyone else. At the end of every flight she would do her make-up before emerging from the cockpit looking effortlessly glamorous.

FIRST WOMAN TO BREAK THE SOUND BARRIER

As aviation evolved, planes flew at greater speeds and distances, but it was not known if an aeroplane, or indeed humans, would be able to travel faster than the speed of sound and break the sound barrier. Some thought the turbulence and air pressure would cause a plane to break up.

The US military developed the Bell X-1, a rocket-powered test plane. It didn't fly like a normal plane, but was dropped from underneath a heavy bomber, fired its rocket engines to accelerate and then glided back to land. Chuck Yeager flew it to become the first person ever to break the sound barrier.

Jackie, determined to set as many records as she could, set her eyes on the prize and sought out Yeager for advice. In 1953 she became the first woman to break the sound barrier – at over 650 mph in an F-86 Sabre jet borrowed from the Royal Canadian Air Force.

Jackie was a hugely successful and accomplished pilot. During the Second World War, inspired by efforts in Great Britain, she campaigned to give female pilots a role, and was the director of the Women's Airforce Service Pilots (WASPs). Her successful cosmetics company rivalled established names like Helena Rubinstein and Elizabeth Arden. Jackie's resolve and ambition saw her push herself to break as many barriers as possible, setting numerous records and leaving a lasting legacy.

'I might have been born in a hovel but I am determined to travel with the wind and the stars.'

THE DAWN OF THE SPACE AGE

THE DAWN OF THE SPACE AGE
1957 → 1972

The space age started as a race. The USA and the Soviet Union, both using technology developed in Germany during the Second World War, wanted to prove they were the most powerful country in the world by conquering space. The Soviet Union made history on 4 October 1957 when Sputnik, the first satellite, orbited the Earth. It emitted a radio signal with a 'beep beep' that could be heard by anyone in the world who tuned in. When the Soviets put the first person in space, Yuri Gagarin, in 1961, the USA responded by throwing down the gauntlet – the goal of putting a human on the surface of the Moon before the end of the decade and returning them safely to Earth.

VALENTINA TERESHKOVA

PARACHUTIST   COSMONAUT

RUSSIA   BORN 1937 →

Valentina Tereshkova wanted to be a train driver. She saw them on their way into Moscow and thought it was the best job in the world. What she actually went on to do was the best job out of this world.

The early days of human space flight were a competition between the United States of America and the Soviet Union. The Soviet Union had put the first person in space, a man named Yuri Gagarin, and now they wanted to put the first woman there. The spacecraft they had built could not land softly back on Earth, so anyone flying in it had to fire an ejector seat at 7 kilometres above the surface and parachute down.

As a young girl Valentina was desperate for adventure and had joined a parachuting club, where she quickly excelled. When the authorities were looking for talented females to become cosmonauts and fly their spacecraft, Valentina was selected.

The mission programme was carried out in secret and she couldn't even tell her mother about it. Instead, she told her she was training for the National Parachuting Team, and wrote letters home making up all sorts of stories.

FIRST WOMAN IN SPACE

Valentina blasted into space on a momentous day in June 1963 and spent three days orbiting the Earth. When she was launched, her mother, Elena, still had no idea that her daughter was a cosmonaut. It was not until Elena's neighbours rushed round and she saw the images of Valentina on television that she realised her daughter was the first woman in space.

Valentina's flight almost ended in disaster. One of the engineers who built the spacecraft made an error in loading the software which controlled the braking system, and it was set to send Valentina far out into space, rather than back to Earth. Fortunately, Valentina spotted the mistake and told mission control, who were able to fix the problem just in time. However, they did not want to admit to their mistake, and for thirty years Valentina kept the secret and took the blame for many of the flight's problems. Eventually, the engineer who had made the error decided that the world should know the truth.

Valentina still dreams of space and has said that she would take a one-way trip to Mars. She will for ever be known as the first woman in space, who paved the way for the hopes and dreams of so many who have followed in her footsteps.

'On Earth, men and women are taking the same risks. Why shouldn't we be taking the same risks in space?'

JERRIE COBB

 PILOT

USA BORN 1931 →

Jerrie Cobb has always been fiercely determined. Incredibly talented, and one of a small handful of female pilots in the 1950s, she has fought through her life to prove that women are as able as men, both physically and mentally.

When NASA were selecting their first astronauts for the Mercury programme, scientists and doctors did not know exactly what would happen to the human body in space. NASA approached a doctor, Randy Lovelace, and asked him to find the most perfect physical specimens. He put some pilots through an extraordinary series of physical tests, possibly the most thorough medical ever. After selection, NASA announced their first seven astronauts, known as the Mercury 7 – all white, male test pilots.

Lovelace was interested to know how women might fare under the same tests. Some people thought women would make better astronauts, being generally smaller and lighter (and therefore easier to launch), requiring fewer resources to stay alive and possibly being better at staying calm under pressure.

In September 1959, by sheer chance, Jerrie and a colleague were walking on the beach as Lovelace emerged from the sea after an early-morning swim, and Lovelace and Jerrie were introduced. Lovelace found out that Jerrie had been a pilot for sixteen years and that she excelled at flying and he knew that he'd found the person he'd been looking for.

Lovelace asked Jerrie if she would be prepared to undergo the same tests as the Mercury astronauts and she jumped at the chance. She suggested some names of other female pilots, and between them they found a group of nineteen women who would undergo the examinations.

Jerrie went through the tests first, and passed with flying colours. Many of the women scored better on the tests than the Mercury 7, and the doctors observed that they all complained far less as they went through the physical onslaught of needles, X-rays and electrodes.

Sadly, though, the programme was suddenly stopped. It was not a NASA programme and the navy refused to allow Lovelace further access to their facilities. Jerrie and others lobbied the American courts, but the conclusion was that women weren't needed. NASA were focusing on the race to the Moon.

Jerrie was hired by NASA as a consultant but after three years of feeling not very consulted, she left and moved to the Amazonian jungle to work as a missionary and solo pilot, delivering food, medicine and other aid to the indigenous people in the rainforest. Forty years later she still harbours a burning desire to go into space – it is the one thing that would have brought her back from the jungle.

DO WOMEN MAKE BETTER ASTRONAUTS?

THE MERCURY 7 WIVES

USA   1950s & 1960s

When NASA revealed which seven astronauts would be the first Americans to venture into the unknown frontiers of space, they gave them the company's full support. But their wives – **RENE CARPENTER, TRUDY COOPER, ANNIE GLENN, BETTY GRISSOM, JO SCHIRRA, LOUISE SHEPARD** and **MARGE SLAYTON** – were left to fend for themselves, unprepared for what was about to happen. Their tremendous stability and support made a vital contribution to the early space programme.

The astronauts, known collectively as the Mercury 7, became instant celebrities when introduced to the world's press on 9 April 1959. Journalists immediately began to clamour for information about their wives and families, knocking on their home front doors, thrusting cameras and microphones at the unsuspecting occupants.

The wives of the Mercury 7 faced great pressure to present a perfect image to the outside world: model housewives for women across America to live up to. All this as their husbands were spending weeks away from home training and risking their lives on brand-new rocket technology, and whilst they were anxiously worrying about whether the men would make it safely back to Earth.

These women coped by coming together, supporting each other through each mission and the unique challenges of their situation. Whenever the press cornered them after a launch, they would always respond that they felt 'proud, happy and thrilled'.

As the astronaut corps has grown through the years since, so has the community of families. Astronaut training takes place all around the world, as the crew learn about the different equipment and experiments which various countries have built and contributed to the International Space Station (ISS). Astronauts can be away from their families for long periods of time even when on Earth.

The flights of the Mercury astronauts only lasted for a few hours, but today missions usually last for many months. When in space, astronauts have lots of ways of communicating with friends and family back on Earth – email and telephone links, regular video conferences – so they can remain closely in touch. Families send small Christmas and birthday presents up to space on cargo ships, as well as other surprises and personal items.

The wives of the first astronauts were unexpectedly thrust into the limelight but came together to support each other and their partners as they ventured into new frontiers. The community of friends and family is just as strong today, providing support to one another whether in space or on Earth, just as anyone would to their own families.

'We really didn't realise what his first flight was going to cause – that the whole country would come together and be so excited.'

ANNIE GLENN

EILENE GALLOWAY

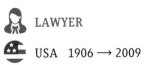

LAWYER

USA 1906 → 2009

Eilene Galloway was passionate that space should be used for peaceful purposes, and spent her long career working to unify people to this end. Throughout her life she would work in areas that no one had ever thought about before.

The early days of the space age were a competition between the United States of America and the Soviet Union. When the Soviet Union put the first satellite in space, many were worried it might start a conflict that could turn into war.

At the time, Eilene was National Defence Analyst and an expert in missiles. A top government official asked for her help in organising meetings to discuss the threat of space. Eilene agreed, even though all she knew about space at the time was that 'the cow had jumped over the Moon!'

She and the government assumed that the race to conquer space was a military one, a question of technology and rocket power. However, scientists and engineers at the meetings explained its enormous potential. There were lots of practical applications, such as using satellites to observe the Earth and monitor the weather, as well as communications, navigation and exploration of the Moon and planets. The benefits to humankind could be tremendous. Eilene and her colleagues saw that space shouldn't be about war, but used for the good of everyone, and in order to keep the peace they would need to devise rules and regulations for space.

LAYING THE FOUNDATIONS OF SPACE LAW

Eilene was asked to produce documents that would first of all create NASA, the American space agency, and then start writing the necessary laws. The USA took their ideas to the United Nations, which then set up a committee to look at peace in space.

Over the next decade, Eilene and her colleagues worked with many different countries to draft new legislation. Even though only the USA and Soviet Union had actually made it to space, many other countries wanted to ensure that it remained peaceful, in case they wanted to make use of it in the future. Finally, in 1967, the United Nations unanimously passed the Outer Space Treaty, which laid the foundations for all of space law. It includes wording that the use of outer space shall be peaceful and for the benefit of everyone.

Throughout the rest of her long career Eilene made tremendous contributions to the field of space law, thinking and writing about new issues like the commercialisation of space. Her passion meant that the world continues to aim to use space for the good of all humankind.

'I think it's very important to not be too serious all of the time.'

MARY JACKSON

 AEROSPACE ENGINEER

USA 1921 → 2005

Mary Jackson lived in Virginia, USA, at a time when there were unjust racial segregation laws and white people considered black people to be inferior. Mary spent her life fighting this inequality, and helping others to see that they could do so as well.

She was very bright, and pushed herself through school and university to get top marks and a degree in maths and physics. She joined NASA's Langley Research Center in 1951 as a 'computer', doing mathematical calculations for the engineers. Mary worked in the team that used the Supersonic Wind Tunnel, a machine that blasted models of aeroplanes and spacecraft with air moving at nearly twice the speed of sound, so that engineers could test their designs at high speeds.

One day, John Becker, a very senior manager, gave her a calculation. She finished the assignment and handed it in, confident that her work was correct. When Becker questioned the answer, she stood firm; she knew she had done the sums correctly. Back and forth they went but Mary was steadfast. Eventually Becker realised that the inputs he had given her were wrong, and her calculations were perfect. He apologised. Word spread that Mary was a force to be reckoned with, unafraid to stand her ground with top management.

Kaz Czarnecki, Mary's boss at the wind tunnel, could see that she was extremely talented. At the time, most universities did not accept women on engineering courses, and black female engineers were unheard of, but he suggested that she take the exams. Mary thought this was a brilliant idea. She had to take some evening classes at the local university to qualify, and because she was black she needed a special permit to attend. Undaunted, she applied for the permit, got the qualifications, was promoted, and in 1958 became NASA's first black female engineer.

NASA'S FIRST BLACK FEMALE ENGINEER

Mary worked in the wind tunnel group for twenty years, producing lots of valuable research and slowly earning promotions. In fact, she stayed there for so long that her hearing was damaged by the loud noise. Eventually she made a tough decision – she left her beloved work in the wind tunnel and took a demotion to become the Federal Women's Program Manager, working tirelessly to improve the lives and careers of all female employees at Langley. She wanted to make sure that there would be many more women to follow in her groundbreaking footsteps.

'We have to do something ... to get [young people]
interested in science ... Sometimes they are not
aware of black scientists, and don't even know of
the career opportunities until it is too late.'

DEE O'HARA

NURSE

USA BORN 1935 →

Dee O'Hara became a nurse because she liked working with people. One day her room-mate said, 'Let's join the Air Force and see the world,' and so they both walked into the recruitment office and did just that. Dee found herself seeing much more than just our world, and had a front-row seat to the earliest days of the space race.

She had been stationed at Patrick Air Force Base in Florida for just a few months when she was summoned to the commander's office. She was terrified, thinking she must be in trouble, but instead he asked her if she would go to Cape Canaveral as nurse for the Mercury astronauts, who had just been selected. Dee didn't know what an astronaut was and had never even heard of NASA, but she took the job.

There had never been a nurse for astronauts before, so Dee had to learn as she worked. She set up the aeromedical lab, with the pre-flight medical area, crew sleeping quarters, conference rooms and labs. For each mission, she took care of the pre-flight medical checks, taking urine and blood samples, and generally looked after the crew's physical health. The first time Dee met the astronauts she was terrified – they were all big celebrities – but they made her welcome and soon became her friends. Astronauts often had an uneasy relationship with flight doctors as they could ground the crew members, so they would tend to hide health issues from them. Dee made a pact with them that if, medically and ethically, she had to tell the doctors something she would, but otherwise they could trust that nothing they told her would go any further.

CARING FOR THE ASTRONAUTS, TO KEEP THEM FIT AND WELL IN SPACE

Dee was a constant presence in the astronauts' lives, looking after their families' health as well. She worked through all the Mercury, Gemini, Apollo and Skylab missions, but never found launch day any easier. She was always apprehensive when the crew were on top of a giant firework, even though she trusted all the machinery to work.

In 1973 Dee left the world of launch operations and moved to NASA's Ames Research Center, to run the human research facility. She coordinated experiments that used bed rest to simulate weightlessness, to learn more about how the human body responds to long periods in space.

To this day she will not reveal any of the secrets the astronauts entrusted to her.

KATHERINE JOHNSON

$\frac{x}{\sqrt{\pi}}$ MATHEMATICIAN ⚛ PHYSICIST

USA BORN 1918 →

Katherine Johnson's extraordinary intelligence, fierce willpower, and determination never to be defined by gender or race led her to some truly phenomenal achievements.

She always loved numbers. Her teachers recognised her great promise early on, and she left school for university at fourteen, getting a maths degree at just eighteen – an achievement even more amazing at a time when education for most black girls finished before the end of school. Her professor said, 'I think you'd be a great research mathematician.' She asked, 'What's that?' and he replied, 'That's for you to find out,' and left it at that.

Katherine became a teacher, got married and raised a family, but she never forgot her professor's words. One day a friend told her about a government research centre that was hiring mathematicians. Katherine applied, and was overjoyed to find herself offered her dream job as a research mathematician.

Engineers would give her reams of data and she loved working through every line, solving equations like a human computer. When assigned to the Flight Research Division, she asked, always inquisitive, if she could attend briefings with the engineers. 'Women don't go to those,' she was told. 'Is there a law?' she fired back. 'Well, no...' came the reply. So she went.

Soon she was calculating flight paths for the Mercury spacecraft, working out exactly

CALCULATING THE PATHS OF SPACECRAFT

when Alan Shepard, America's first astronaut in space, had to launch and land. During the flight she was very nervous, and relieved that her calculations were correct as he came safely back.

Later, John Glenn's flight was the first American astronaut flight planned to orbit the Earth, and the trajectory was even more complicated. Although electronic computers had been introduced, Glenn insisted that Katherine check all the numbers. 'If she says it's right, it's right.' She went on to calculate the trajectories for the Apollo flights, getting humans to the Moon for the very first time.

Katherine worked at NASA until her retirement in 1986. She always did her best on everything she worked on, so she never had to be asked to do it a second time.

Over the years she has received numerous awards, including the Presidential Medal of Freedom, but some of her most treasured mementos are the letters from young schoolchildren, inspired by her work and the way she broke down race and gender barriers. Her love of numbers and dedication to her work meant that she went from counting everything around her to changing everything around her.

'If you want to know, you ask the question.
There's no such thing as a dumb question. It's
dumb if you don't ask it.'

MARGARET HAMILTON

1001
0101 SOFTWARE ENGINEER
1010

 USA BORN 1936 →

Margaret Hamilton is one of the first pioneers of software engineering. When she started working with computers the job didn't even exist – she coined the term 'software engineering' to describe her work.

Margaret loved maths and went to university to study the subject, working on computers so huge that some took up whole rooms. Whilst there, she heard about the job of a lifetime. Massachusetts Institute of Technology (MIT), who were working for NASA to develop the software to send people to the Moon, were looking for new recruits. Margaret rushed to call MIT on the phone and within hours had set up interviews with two project managers. Both wanted to hire her, so she told them to flip a coin, secretly hoping her favourite would win. Fortunately he did.

Margaret was in charge of the team that wrote the on-board software for the Apollo Lunar Module and Command Module, the spacecraft that would take American astronauts to the Moon. It was a very steep learning curve for them all as no one had done this sort of thing before. They knew that there were no second chances for their software – it needed to work perfectly – and so they took the work very seriously.

DEVELOPING SOFTWARE TO KEEP SPACE MISSIONS ON TRACK

Margaret tried to spend as much time as possible with her young daughter, Lauren, during this busy period by taking her to work in the evenings and at weekends. One day Lauren was playing with a test version of the computer and accidentally crashed the software. Margaret wanted to fix the code to prevent astronauts from doing the same thing, but NASA would not let her, saying they wouldn't make the same mistake as a young child. But during the Apollo 8 mission, astronaut Jim Lovell inadvertently did exactly that and accidentally wiped some data. Margaret and her team had to fix the problem while the crew were in space.

All the long hours and hard work paid off. The software they had developed was so robust that no bugs were ever found. Even when Neil Armstrong and Buzz Aldrin were on their final descent to the Moon and the computer became overloaded, it was able to prioritise its functions so the crew could land safely. The software formed the basis of that later used in Skylab and the Space Shuttle, as well as the digital systems in aeroplanes. Margaret's pioneering work continues to keep humans reaching for the stars.

'Only those who dare to fail greatly
can ever achieve greatly.'

THE WALTHAM 'LITTLE OLD LADIES'

 TEXTILE WORKERS   WATCHMAKERS

USA 1960s

The Apollo Guidance Computer (AGC) was groundbreaking but even so, it was nothing like the computers of today. We are used to handheld smartphones and tablets on which programs can easily be installed and uninstalled. When the AGC was being designed in the 1960s, computer disks were extremely fragile. There was no way they would survive the vibrations and g-forces of a rocket launch.

Instead, the engineers who designed the computer used 'rope memory'. Computer code language is made up of 1s and 0s, which when written into 'words' will tell the computer what to do. Rope memory is made up of rings and fine copper wires – a wire going through a ring represents a 1, and one going around the outside a 0, together writing the instructions for the computer.

When the engineers looked at their designs, they saw that the AGC would require thousands of detailed and painstakingly handmade parts. They turned to the women of Waltham, Massachusetts, experts in textile weaving and making watches.

PRECISION WORKERS WEAVING APOLLO'S GUIDANCE COMPUTER MEMORY

To build the rope memory, two women would sit opposite each other in front of a loom. They would read the written computer program and weave it into the rope by passing a needle with the copper wire in it backwards and forwards. The work was painstakingly slow – one program could take several months to weave and errors were time-consuming to correct. The ladies knew that the lives of the astronauts going to the Moon depended on their handiwork, so they did their utmost to produce work of the very best standard.

The engineers who wrote the programs nicknamed this technique the LOL method, after 'Little Old Ladies', though this was a hugely misleading description. These women were so important and skilled that sometimes they were paid just to do nothing, to make sure they weren't busy on another job when they were needed.

This process of making the rope memory was very cumbersome but it meant that it was impossible for the program to be deleted. When Apollo 12 was struck by lightning just thirty-six seconds after lift-off, the sudden electrical discharge through the rocket knocked some systems offline. Thanks to the rope memory, the computers rebooted themselves, mission control responded to the problem and the crew continued on their way to orbit and the Moon.

Sadly, so many of the names of the women who spent hours lovingly building these computers appear, for now, to be lost in time, but their contributions have not been forgotten.

'We used to go to the cafeteria and the astronauts
would come in ... They'd explain the Moon shot
and thank us for what a good job we were doing.'

MARY LOU ROGERS, one of the ladies who worked on the Apollo line

POPPY NORTHCUTT

↓⚙ ENGINEER 👩 LAWYER

🇺🇸 USA BORN 1943 →

Poppy Northcutt has spent her whole life breaking the mould. She studied maths at university, not just because she enjoyed it, but also because she saw it would help her to reach higher-paid roles than those traditionally available to women. After graduating she took her talent for maths to work at an aerospace company. Here she solved very difficult equations for the engineers who were working on software that calculated how to get a spacecraft back from the Moon. Poppy was fascinated by her work, so she asked lots of questions and even took the computer code home with her to study it at night. Before long she was finding mistakes in the code and was promoted to engineer.

Apollo 8 was the first mission that would take human beings to orbit the Moon, a journey of about 240,000 miles and four days. During a flight, astronauts in a spacecraft are supported by a team of engineers in mission control, who monitor everything that is going on. They are a vital part of the mission, sending commands, solving problems, keeping the astronauts safe and well. The computer code that Poppy and her colleagues had been working on was used to calculate the trajectories for the Apollo 8 mission's return, so they were chosen to give operational support. Poppy was the first female engineer to don a headset and work in mission control.

FIRST FEMALE ENGINEER IN MISSION CONTROL

Poppy and her team were rushed in to help when disaster struck the Apollo 13 mission. The oxygen tank exploded, crippling the spacecraft, and the crew could no longer land on the Moon as planned. Poppy and her colleagues had to use their computer programs to quickly work out new trajectories for the spacecraft. Thanks to their lightning-speed calculations they devised a new path for the spacecraft to get back to Earth safely.

Working at NASA, Poppy was aware that she was leading the charge for women into a historically male world. She was passionate about fighting for women's rights and decided to go into politics and law. She changed legislation so that women were allowed to wear trousers to work and weren't treated any differently from men, whether they worked in the police or the fire department. She also went on to become a fantastic criminal defence lawyer. Poppy's time in mission control blazed the way for the many women who have followed in her footsteps, and she continues to fight for women's rights to this day.

'I started looking around at these dudes that were working with me and I thought, "You know, I'm as smart as they are."'

RITA RAPP

⑂ PHYSIOLOGIST 🍖 NUTRITIONIST

🇺🇸 USA 1928 → 1990

Rita Rapp was responsible for one of the most important parts of any mission. She joined NASA in 1961 in the early days of the space programme and started working on the Mercury project, designing in-flight items such as exercise devices and medical packs, but within a few years moved to the Apollo programme to plan what the astronauts would eat in space.

The food eaten by the Mercury astronauts wasn't very tasty – it was mostly purée squeezed out of a tube, or bite-sized cubes of compressed food, covered in gelatine. Food in space floats, just like everything else, which makes crumbs very problematic: instead of falling to the floor they can clog up filters or even get in astronauts' eyes. During the Gemini flights, the food was a little better, but still fairly bland. Gus Grissom smuggled one of his favourite corned-beef sandwiches onto a flight, but it disintegrated as soon as he pulled it out of his pocket!

Rita was responsible for all the food that the astronauts would eat before and during their missions for Apollo, Skylab and the early days of the Shuttle. She was determined to make it taste as good as possible and to give the crews food that was much more like what

SERVING UP SOME FLAVOUR IN ORBIT

they ate when on Earth. She worked hard to improve nutrition, to design and develop new food packaging systems, and even add cutlery. Rita would also bake home-made bite-size cookies for the crew. These were so tasty that during the Skylab missions they were used as on-board currency, with personal cookie allocations being traded for favours.

Space food today is not dissimilar from what we eat at home. In order to reduce weight, and therefore launch costs, much is sent in a dehydrated state, like instant mashed potato. The astronauts simply add hot or cold water, wait a few minutes, then cut open the plastic pouch and eat it. Other food comes in tins or metallic pouches, and is warmed in a device that is a cross between a suitcase and a sandwich grill. Things like chocolate, nuts or biscuits go just as they are.

Rita knew that if astronauts had food they liked, they would work better in space. They were all very grateful for her pioneering efforts, transforming their meals from 'cubes and tubes' to the space version of haute cuisine.

'You can't just go to the grocery and put things on board. It's a challenge, but I enjoy meeting the challenge and accomplishing the end.'

DOTTIE LEE

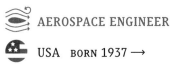

AEROSPACE ENGINEER

USA BORN 1937 →

Dottie Lee was there at the beginning of space travel. A natural mathematician, she studied maths at university with plans to become a teacher but instead her career went on to take a galactic route.

In 1948 she started at NACA, later to become NASA, as a 'computer', a term that meant something different then to today: someone who supported the engineers, working out sums on huge mechanical calculators. One day she was asked to cover for a secretary and in between answering phone calls she solved an incredibly difficult calculation. The secretary's boss was so impressed that he asked her to join his engineering team. Dottie learnt everything she could, working on more and more complicated projects, eventually becoming an expert in aerodynamics and heat shields.

When spacecraft come back to Earth, they have to slow down. One way to do this is to fire up the rocket engines for a long time, but this uses a lot of fuel. Instead, engineers choose to use the drag of the atmosphere to slow the spacecraft. A capsule coming back from a low orbit does fire its engines for a few minutes, slowing it to about 8 kilometres per second, but the Apollo flight coming back from the Moon just aimed straight at the Earth, re-entering the atmosphere at about 11 kilometres per second. When a spacecraft re-enters the atmosphere it hits the air molecules with great force, causing the surfaces to heat up to very high temperatures, over 1,600 degrees Celsius. If an object has no protection, it will burn up in the atmosphere, like a meteorite, so spacecraft bringing humans or other cargo back to Earth have a protective surface, called a heat shield. Dottie and her team calculated the amount of heat that the Apollo spacecraft would have to withstand and made sure that the design would protect the crew inside.

Dottie was always a step ahead – when Neil Armstrong and Buzz Aldrin were walking on the Moon, she was already in meetings starting to design NASA's next spacecraft, the Space

CRUNCHING THE NUMBERS FOR APOLLO'S HEAT SHIELD

Shuttle. She became responsible for making sure it didn't overheat on the way to or from space, and for designing the heat shield, which on the Shuttle was lots of clever ceramic tiles. The nose of the Shuttle became known as 'Dottie's nose' after she came up with the best shape.

Dottie worked for NASA for her entire career, always enthusiastic about her work, and a role model for all. She was so good at her job that when she retired they had to hire ten people to cover all the work that she used to do.

'You learn with each experience, of course, and that's
what I did every day of my life.'

THE ILC SEAMSTRESSES

 SEAMSTRESSES

USA 1960s

ELLIE FORAKER, **MADELEINE IVORY**, **BERT PILKENTON** and **CECIL WEBB** were among the expert seamstresses who made women's underwear for the International Latex Company (ILC). It came as something of a surprise to them when they were asked to use their talents making possibly the most complex garment ever created.

When NASA wanted to make the spacesuit for the Apollo moonwalkers, they needed a design strong enough to survive the extremes of space, but which also let the astronauts move. They turned to ILC, who came up with a game-changing design. It included a brilliant piece of engineering for the joints, making them like a concertina – an idea now familiar in bendy drinking straws but revolutionary at the time.

ILC custom-made three spacesuits for each astronaut – one for training, one for flight and one as an emergency backup. Each was a mini spaceship, providing air to breathe, communications with the crew, and even a nappy just in case. To achieve all this, each suit was made of twenty-one different layers, sewn one inside the other.

These mind-blowingly intricate suits were all made by hand, using regular sewing machines. Pins were strictly rationed as any forgotten ones could puncture the layers and cause disaster. The workload was huge and some teams worked around the clock to get the job done. They all knew that their handiwork was the only thing protecting the astronauts on the moonwalks, so they did their very best and the astronauts really appreciated it. Jim Lovell, the commander of Apollo 13, wrote them a note one day: 'Thank you for sewing straight and careful. I would hate to have a tear in my pants while on the Moon.' The women were incredibly proud of their work and left their names inside the suits.

When Neil Armstrong and Buzz Aldrin landed on the Moon and ventured out onto the surface, taking humankind's first footsteps on another world, everyone at ILC was holding their breath. It was the first time that their work had ever been used in space and they watched anxiously as the two men bent and flexed, stumbling and falling as they learnt how to walk in the Moon's gravity. They needn't have worried; their dedication and skill had produced brilliant spacesuits – flexible and tough, and perfect for working on the Moon.

Ellie, Ruth and others went on to make suits for the Space Shuttle spacewalks and parachutes for Mars rovers. ILC are currently developing new suits that one day might be worn on Mars. From bras to Mars, the skill of these women has kept astronauts alive and humanity landing safely on new worlds.

MADE-TO-MEASURE SPACESUITS

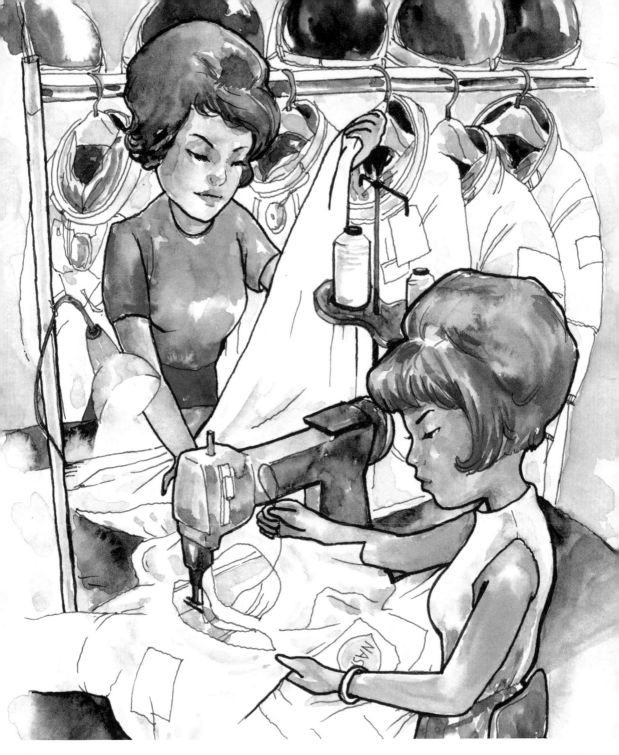

'I had something to do that was great. I did something great in my lifetime. I built the suit that went to the Moon.'

BERT PILKENTON

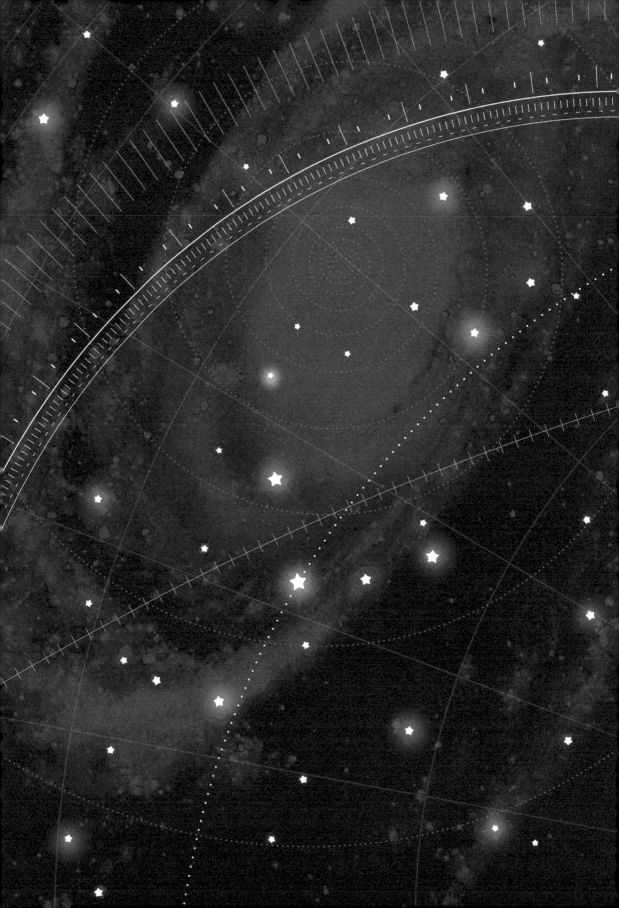

SPACE
STATIONS
AND
SHUTTLES

SPACE STATIONS AND SHUTTLES
1972 —→ 2000

After the last footprints on the Moon were made in 1972, the USA and the Soviet Union turned their attention to new goals. They started to work together in space, rather than competing. The Soviets focused on space stations, starting out with building the Salyut spacecraft, and then the first modular space station, Mir. The USA developed the first reusable spacecraft, their fleet of Space Shuttles, which would haul satellites, telescopes and scientific instruments into space. Both began inviting people from other countries to fly with them and the world of space travel became an international endeavour.

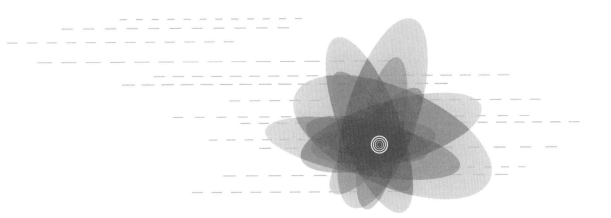

SALLY RIDE

 PHYSICIST   ASTRONAUT

USA   1951 ⟶ 2012

Sally Ride was an intensely private individual, an introvert who liked to keep herself to herself. But her drive, her skill and her intelligence meant that she went on to be known around the world.

FIRST AMERICAN WOMAN IN SPACE

Growing up, Sally was a brilliantly talented tennis player but she also loved maths and science, and particularly physics. She could have successfully followed either passion but decided that science would be the better option for her in the long term. She was in the final months of completing her PhD in astrophysics when she saw an advert in her university newspaper recruiting for astronauts and, for the first time, encouraging women to apply. Sally thought to herself, 'I could do that – it sounds like fun.'

In 1978 she was hired by NASA as an astronaut, and along with thirty-five other people trained to fly the new Space Shuttle. There were five other women: Anna Fisher, Shannon Lucid, Judy Resnik, Rhea Seddon and Kathryn Sullivan. NASA was almost completely male at the time, so they had to show their colleagues that women are as capable as men. Slowly, over time, the talent and drive of the new recruits shone through and it became clear that it didn't matter if you were male or female.

Sally had great knowledge and skill in operating the Shuttle's robotic arm and was selected to fly on the fifth Shuttle mission. When the crew of seven blasted off in June 1983, the world's media was focused on the first female American astronaut, though Sally wanted just to be part of the team. The mission went very smoothly, the whole crew playing their part; successfully launching two satellites and carrying out lots of experiments before returning to Earth after six days. When she came back, Sally saw what an impact her flight had had on people and she realised what an inspiration she had been, in particular to young people all around the world. Sally resolved to keep helping them follow their dreams in science and technology.

Sally worked tirelessly to encourage children to see what excitement and wonder lay in the sciences and her name is taught in schools worldwide. She blazed the trail, not just for her fellow female astronauts, but for women all over the world, showing that they could and should work as equals to men, both on and off the planet.

'There was absolutely no sense – through all the years of growing up – that there was any limit to what I could do or what pursue.'

SVETLANA SAVITSKAYA

PILOT COSMONAUT

RUSSIA BORN 1948 →

Svetlana always had flying in her blood. She loved the feeling of moving through the air at great height and speed and this thrill for adventure took her to flights way above the Earth.

In the early days of the space race the Soviet Union and the USA were locked in a battle to conquer the cosmos. Although the culture in the Soviet Union at the time saw a woman's role as being in the home, the Soviets took tremendous pride in securing the historical first flights, including those by women, and went to great lengths to upstage the Americans. By 1978, the USA were ahead in the race with the achievements of their Apollo programme and the Moon landings. Now they were including women among their crew, announcing Sally Ride as their first female astronaut.

In response the Soviet Union trained female cosmonauts of their own. Svetlana was a pilot and parachutist, had flown MIGs and was the world aerobatic champion, so in their search for potential cosmonauts her expertise and daring were a perfect match. When the USA announced Sally Ride's flight, the USSR assigned Svetlana to a flight that was visiting Salyut 7 space station, which would launch before Sally did. As soon as she floated on board the station one of her crew mates joked that as a woman she should get in the kitchen. She coolly replied, 'I thought you would be the one to fix us something to eat.' She was unflappable in the face of these sexist views and went about her duties with great skill, performing a number of scientific and medical experiments and proving herself a high-class cosmonaut.

Two years later, the Soviet Union again wanted to upstage the American space programme so they sent the talented Svetlana on another mission. She became the first woman to venture outside the safety of her craft and make a spacewalk, launching herself into the unknown and testing important new tools.

FIRST
WOMAN
SPACEWALKER

Spacewalking, or Extra Vehicular Activity (EVA), is a very important part of living and working in space. Astronauts and cosmonauts need to be able to get outside to fix malfunctioning parts, to set up new experiments or to attach new modules to the space station. A spacewalk usually lasts six to eight hours and the spacesuits are mini-spacecraft, keeping them alive, protecting them from the extreme temperatures of space, providing air to breathe and communications equipment as they step into the void.

Svetlana's achievements should never be diminished because of the Soviet state's attitude towards women – they should instead be celebrated and are a reminder to all that women can achieve anything.

'Women go into space because they measure up to the job. They can do it.'

NICHELLE NICHOLS

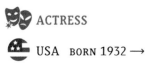 ACTRESS

USA BORN 1932 →

Nichelle Nichols is a sci-fi icon. She kicked down boundaries when playing the strong character of Lieutenant Uhura in the original 1960s series of *Star Trek*. She became a role model across the world, showing that black people could take leading parts on both stage and screen, sharing the first interracial TV kiss and helping to portray a future world that was free from sexism and racism. What is less well known is that she also helped kick down the barriers to real-life space travel.

INSPIRING WOMEN AND MINORITIES TO REACH FOR THE STARS

The first astronauts recruited in the USA were all white, male and mostly from the military. At a *Star Trek* convention in 1975, Nichelle heard the NASA Director of Sciences speak. She realised that the space programme was not representative of the world at large, and she felt very strongly that astronauts should come from all walks of life. During a speech at the National Space Institute she made this point and NASA started to listen. Soon afterwards they asked her to help with their ongoing astronaut recruitment campaign, which, a few months in, still had very few applications from women and minority groups. Nichelle was determined to find the very best female and minority candidates for NASA and made it clear that if some of them weren't selected, despite their eminently suitable qualifications, the world would know about it.

Nichelle didn't just make a couple of videos; she immersed herself in the programme, visiting specialist centres and going through parts of astronaut training, all so she could understand the skills and qualities needed to be an astronaut and communicate them to her audiences. At the same time she visited schools and universities across the United States, who welcomed this sci-fi celebrity with open arms and then listened intently to her message that gender and race should be no barrier to space. Her perseverance paid off. The number of applications that NASA received from women and minorities soon soared and when they announced their new class of thirty-five astronauts in 1978, it included six women and four people from minority backgrounds.

Nichelle didn't sit back and relax after the first success, though. She has continued to work closely with NASA to the present day, highlighting the amazing work that is done throughout the world of space. She also encourages all people, whatever their background, to boldly go where no one, man or woman, has gone before.

'Science is not a boy's game, it's not a girl's
game. It's everyone's game. It's about where
we are and where we're going.'

CHRISTA McAULIFFE   JUDY RESNIK

TEACHER   ASTRONAUT   ELECTRICAL ENGINEER   ASTRONAUT

USA   1948 → 1986   USA   1949 → 1986

When the twenty-fifth Shuttle flight, STS-51-L, roared away from the launch pad on a bright and cold morning in 1986, much of the world considered space flight routine. The events that followed reminded everyone that this was not the case.

NASA's fleet of Space Shuttles had been blasting into space every few weeks throughout 1985. On this occasion, children around the world had tuned in to see Christa McAuliffe head up to teach the first ever school lesson from space. She was accompanied on board Challenger by six other crew mates: the second American woman ever to fly to space, Judy Resnik, alongside Gregory Jarvis, Ronald McNair, Ellison Onizuka, Francis Scobee and Michael Smith.

Judy loved engineering. She refused to listen to those who thought women couldn't be engineers and worked hard at everything she put her mind to, including maths and playing the piano. Judy was one of the first female astronauts recruited by NASA, and the first Jewish woman in space, but labels weren't important. What really mattered was to do her very best.

Christa wasn't a professional astronaut but a history teacher who was picked by NASA from over 10,000 applicants to give lessons to hundreds of thousands of schoolchildren, live from the Space Shuttle. She had just six months of basic training before flying, compared to the initial year that astronauts must undertake before going on to specific mission training. Christa was really excited about this novel way of engaging students.

On the morning of 28 January 1986 in Florida, temperatures were hovering around freezing, which was the coldest weather that a Shuttle had ever been launched in. Some of the engineers said it wasn't safe to fly, but the programme managers decided to go ahead and the fate of the crew was sealed.

The Space Shuttle was propelled into space by three large engines, and two additional solid rocket boosters that were strapped to the sides. The cold weather had hardened the rubber

THE CHALLENGER ACCIDENT

seals on the boosters, so they didn't bend properly as the shuttle blasted off. Hot exhaust gas leaked out of a joint and just seventy-three seconds after lift-off the external fuel tank exploded, tearing Challenger apart. All the crew were killed.

Space travel will always have risks. We honour Christa and Judy, and all the others who have lost their lives, by not giving up on the challenges of exploring space, building on their work, and opening up new and incredible opportunities for all of us on Earth.

'People who you consider to be heroes are really quite like yourselves. Only hard work and perseverance will help you to succeed at any venture.'

JUDY RESNIK

'Reach for it.' Push yourself as far as you can.'

CHRISTA McAULIFFE

MAE JEMISON

 DOCTOR   ASTRONAUT

USA   BORN 1956 →

Mae Jemison grew up in a whirlwind of creativity and imagination. Inspired by the Apollo missions and science fiction such as *Star Trek*, she always assumed she would go into space one day. She was a brilliant dancer, loved fashion design and art, but was also great at science so she studied medicine and became a doctor. When she had the chance to apply to be an astronaut, she decided to give it a shot. As she was making her application, it never crossed her mind that no black woman had ever been to space before. Mae was selected by NASA and spent eight days in space on STS-47 in September 1992.

FIRST BLACK WOMAN IN SPACE

During her mission Mae worked with a number of experiments, including producing the first non-insect babies conceived and hatched in space. She fertilised some frog's eggs, and then observed as the spawn hatched into tadpoles, which grew in space and came back to Earth. She paid homage to one of her great inspirations, the actress and activist Nichelle Nichols, by starting each of her shifts calling down to Earth, 'Hailing frequencies open,' Lieutenant Uhura's famous line from *Star Trek*.

Mae wanted to use her platform as the first black woman in space to help inspire women of all races to be involved with science, helping to shape the development of the world. She left NASA and set up a foundation named in honour of her mother, Dorothy, to promote science and technology. She says that her parents were the best scientists she knew, because they were always asking questions. Mae also brought the world of space and sci-fi together when she appeared in an episode of *Star Trek: The Next Generation*. This made her the only real-life astronaut to have also served on the Starship Enterprise.

Continuing to be even more ambitious, Mae set up an initiative called 100 Year Starship. This encourages people to think big, and to nurture the giant leaps in knowledge that will enable humans to travel beyond our solar system to another star within the next hundred years. Such a project sounds like a crazy idea to some, but Mae looks back to when, in 1901, the idea of a person walking on the Moon seemed equally unthinkable, though by 1969 that is what had happened. Mae's vision and ambition may turn out not to be science fiction.

BIG PLANS FOR HUMANKIND'S FUTURE IN THE STARS

'Never be limited by other people's limited imaginations.'

HELEN SHARMAN

CHEMIST   ASTRONAUT

UNITED KINGDOM   BORN 1963 →

Helen Sharman describes herself as having a very ordinary upbringing, but one day, listening to the radio in the car, her life started to become rather extraordinary.

In June 1989, when she was a young chemist working in chocolate research, Helen was driving home from work when she heard an advertisement on the radio: 'Astronaut wanted: no experience necessary. Must be physically fit, aged twenty-one to forty, practical, with a scientific degree and a proven ability to learn a foreign language.' 'That's me!' Helen thought, and scribbled the phone number on an old receipt.

A few weeks later, she nearly didn't post the application form, thinking that she was far too ordinary and would never be chosen, so she was very surprised when she got a phone call and was offered a place in the selection process.

The mission was a commercial enterprise, with £16 million of funding to be paid to the Soviet Union, raised through sponsorship from British companies. The selection process was filmed for TV – with the ultimate prize of a trip to space. Helen undertook round after round of selection and lots of medical tests with a determined resolve, and eventually made it to the final four candidates. On a TV show filmed at London's Science Museum it was announced that Helen and Tim Mace were the two going to the Soviet Union for eighteen months of training; one person would fly in space, the other would be the backup crew.

FIRST BRITISH PERSON IN SPACE

As Helen was going through her final gruelling training, the project almost failed because of a lack of funding, but thankfully the Soviet Union took on the cost. In February 1991, just a few weeks before the launch, Helen was overjoyed to find out that she had been assigned to the mission and could hardly believe it was true. On a clear day in May 1991, the dream became a reality and she rocketed into space.

She spent eight days in space, visiting the Mir space station and carrying out lots of experiments, some on herself. She grew wheat and potato seedlings, took some snails and a lemon tree into orbit to see their response to space and spoke to schools using amateur radio.

Helen is the first British person ever to go to space, and the first woman to visit Mir. Helen believes that anyone can be an astronaut and that her 'ordinary' qualities – good health, fitness, the ability to get on with other people and to work in a team – are among the best skills for the job. Helen has shown the world that we can all be extraordinary if we put our minds to it.

'Aim high! If you want to be an astronaut ... get a lot of different experiences of all the things that life has to offer. Just enjoy living on planet Earth.'

EILEEN COLLINS

🪖 PILOT 🔲 ASTRONAUT

🔴 USA BORN 1956 →

Eileen Collins didn't have a privileged upbringing – her family had been through some tough times – but she would read book after book about pilots from her local library. She dreamt of becoming a military pilot, even though she'd never set foot on an aeroplane, and saved money from evening and weekend jobs for flying lessons.

As soon as the US Air Force started accepting applications from women, Eileen applied

TRAINING AIR FORCE PILOTS

to join and became an excellent pilot. When she started teaching other people to fly jets, she was the only female doing so in her squadron and was careful to do a great job to prove that women could be just as good as men. She did so well that she became only the second woman to graduate from the test pilot school at Edwards Air Force Base.

Not long after she had started her training, her base was visited by newly selected astronauts, including the first female ones, and Eileen decided that she wanted to fly the Space Shuttle one day. So a few years later, when NASA announced they were recruiting more astronauts, she applied, and again her brilliant piloting skills were recognised and she was hired.

The Space Shuttle looked a bit like an aeroplane, but was more like a rocket-powered glider. It was launched with two big rocket boosters and three main engines. At the end of the mission, the Shuttle would fire its main engines once more to re-enter the Earth's atmosphere. From that point on it was a glider; there was no more power and it just used the wings and flaps to control its path as it flew 'S' shapes through the air to slow down. Most gliders are lightweight with long wings, to provide as much lift as

FIRST FEMALE SHUTTLE PILOT AND COMMANDER

possible, but the Shuttle was heavy with a small delta wing so it was often known as a 'flying brick'. The pilot had to be incredibly skilled as there was only one shot at landing – without power there was no chance to go around again.

In February 1995 the Space Shuttle Discovery roared into space, and for the first time ever there was a woman in the pilot's seat – Eileen. The flight was the first Shuttle mission to rendezvous with the Mir space station and she piloted it with exemplary skill and precision. In July 1999, Eileen became the first, and only, woman to take command of a Shuttle mission. Her three-year-old daughter, Bridget, thought that all mums flew spacecraft. Eileen has shown the world that one day many more will.

'Don't miss the opportunities.
Be on the lookout for them.'

CHIAKI MUKAI

DOCTOR   ASTRONAUT

JAPAN BORN 1952 →

Chiaki Mukai was determined to become a doctor. She had seen her brother, who had brittle bones in his legs, suffer and struggle to walk and she wanted to be able to help others like him. Chiaki chased her ambitions and qualified as a doctor. She has spent her life working to help others, but much of her research has happened in places she did not expect.

One day she was reading the newspaper after a night shift at the hospital when she saw that the Japanese space agency, JAXA, were looking for astronauts to fly on their Shuttle. Chiaki shouted out loud, 'Can someone from Japan actually fly in space?' She thought all astronauts had to be American or Russian and did not know that Japan even had a space agency. Nevertheless, the idea intrigued her and the more she thought about it, the more she wondered what the possibilities of space might offer to her medical research, so she applied.

Chiaki was one of the first three Japanese astronauts selected in 1985 and the only woman. On her first flight, STS-65, in 1994, she put her medical expertise to great use, investigating how the human body responds to time in microgravity. She also researched how space affects aquatic life forms, performing experiments on Japanese killifish and fire belly newts.

HOW DOES SPACE AFFECT THE HUMAN BODY?

She was fascinated by how she felt in space, though her training had led her to expect this. What surprised her the most was how her body felt when she returned to Earth. She said that despite knowing about gravity on Earth, she was not aware of it until she came back from space and could feel the heaviness of her body and noticed how everything was drawn to the ground. She learnt not to take for granted the little things that we barely notice in our lives and which have always been there.

Chiaki returned to space in 1998 on board STS-95, making her the first Japanese person to go to space twice. She was training for a third mission when the Columbia accident happened, and decided it was time to step down as an astronaut and look at other research.

Since then, Chiaki has combined her two passions of medicine and space, managing JAXA's Biomedical Research Office and leading many experiments that have continued to fly to space. Her out-of-this-world medical research continues to help those on Earth, like her brother, who are suffering.

'If you can dream it, you can do it.'

CLAUDIE HAIGNERÉ

DOCTOR   ASTRONAUT   POLITICIAN

FRANCE   BORN 1957 →

Claudie Haigneré enjoyed school when she was a child, working hard and learning as much as she could. She was also very athletic and loved running. Claudie had hoped to study sports at university but she had done so well in school, finishing it early, that she was too young to be accepted on the course she had set her sights on. She decided to study medicine at a different institution instead, which had no issues with her young age. She was a very well-respected researcher when she saw the French Space Agency (Centre national d'études spatiales, CNES) were looking for scientists to fly in space.

Claudie was selected and developed many experiments for CNES, particularly focusing on how the human body adapts to life in space. When she took some of her experiments to the Mir space station in 1996, along with those of other scientists, she was the first French woman in space. She returned in 2001, this time visiting the International Space Station (ISS) to carry out more experiments.

In between the two missions she gained a qualification that no other woman has ever received: Claudie qualified as a Soyuz Return Commander, meaning she could fly the Soyuz spacecraft home in an emergency.

The Russian Soyuz spacecraft is a three-person capsule that has been transporting cosmonauts and astronauts to and from space since 1967 and it is considered to be a very reliable workhorse. Since the Space Shuttle was retired in 2011, it has been the only way for people to get to and from the ISS. The Soyuz that the crew arrive in stays docked to the ISS throughout their mission, so they always have a way to

GETTING BACK TO EARTH IN AN EMERGENCY

get home if there is an emergency. Every morning they check over their spacecraft, make sure everything is OK and print out a set of times which tell them at what points during the day they could leave and head back to Earth if they needed to. It is their vital safety net.

Fortunately, during Claudie's missions, and indeed so far throughout the history of space stations, no one has had to come home in an emergency, but the reliable spacecraft is always there, just in case.

Claudie returned to Earth inspired by her visits to space. She went on to have careers in politics, business, science and research. She has even been given the top order of merit in France, the 'Legion of Honour'.

'Dare in your life. Don't wait to be perfect.
Why not you?'

PATRICIA COWINGS

? PSYCHOLOGIST   🧠 AEROSPACE PSYCHOPHYSIOLOGIST

🇺🇸 USA BORN 1948 →

Patricia Cowings has spent her career trying to help astronauts feel healthy and ready to achieve great things in space.

Her parents encouraged her as a child to do something she was passionate about and she decided to study psychology. When she was at university, the engineering department had a course about how to design the equipment astronauts would use on the Space Shuttle. Patricia didn't have the right qualifications, but she wanted to take the course. Seeing that it was full of men, she told the professor in charge, 'You have to let me in! You need a woman in the class.' It turned out that this would shape Patricia's whole life.

During the course, she visited NASA and learnt about how space flight affected astronauts' bodies. She became interested in how the mind and body interact and how the mind could help solve medical issues, particularly space sickness. Patricia has spent her career developing a technique called Autogenic-Feedback Training Exercise, teaching astronauts and other people how to control some of their physiological responses, such as heart rate and blood pressure, helping them to overcome motion sickness.

INVESTIGATING SPACE SICKNESS

We live with the gravitational pull of the Earth acting on us all day. Our bodies are used to working against it, but when a human goes into space and starts floating, all the fluid inside the body begins to even out. The balance system in our ears relies on fluid, and as we float and the fluid inside our ears floats, the brain is unsure how to deal with the changed signals. The result is feeling sick, a bit like seasickness. Fortunately, the body is very adaptable and after a few days the brain figures out what is happening and learns to ignore the signals and you stop feeling sick. The same thing happens in reverse when you come back to Earth, but it can sometimes take even longer for the brain to figure out the signals again.

PUSHING HUMAN MINDS AND BODIES TO THE LIMIT

Patricia's work has been essential in trying to understand why some people get more space sick than others and in finding ways of helping astronauts adapt to the changes their bodies face when they go to and from space. It has also helped lots of people on Earth with balance problems and could be crucial if humans go to Mars where there will be no one to assist them. Patricia's vital research is helping humans push the boundaries of their bodies and our knowledge of the mind further than ever before.

'Doesn't matter where you are from or what you look like ... I've spent my life studying human potential — and stretching my own.'

IRENE LONG

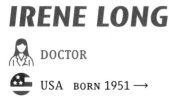

DOCTOR

USA BORN 1951 →

Irene Long's father loved space travel and his enthusiasm inspired his daughter, who would watch him soar through the skies as he learnt to fly aeroplanes. Irene also loved medicine and dreamt of combining her two passions and setting up a clinic on the Moon one day.

Whilst at medical school, she read about an aerospace medicine course. It was somewhere in her home state of Ohio, but she didn't know where. Irene, ever resolute, wrote to numerous addresses, enquiring about the course and how to apply. A few days later she received a phone call. 'Apparently you want to go into space medicine,' said a voice. 'Someone just dumped fifty letters on my desk, and they're all from you!'

The letters worked as Irene became only the second civilian to study on the aerospace medicine programme at the Wright-Patterson Air Force Base. At the end of the course her unwavering ambition shone through again when she told her supervisor, 'If I'm not hired, I'm moving in with you and your wife.' The next day she received a job offer, and has worked with NASA ever since. Irene worked hard and ended up becoming the chief medical officer, responsible for the well-being of everyone at NASA, astronauts and ground crews alike.

The health of an astronaut is hugely important, just as it is for any other person in their job. But being in space brings an added difficulty – if an astronaut becomes ill whilst in space it isn't as simple as popping down to the doctor's surgery. Every astronaut in space has a doctor assigned to look after them, and they have regular conferences from space to make sure all is well. Astronauts are also trained to perform some medical procedures too, from tooth extraction to sewing up wounds, but for any major issues they have to come home.

NASA'S CHIEF MEDICAL OFFICER

Coming home from the International Space Station can be done in a few hours, but the quickest route to Earth means an astronaut might end up in the middle of nowhere fending off bears. The stresses and strains of re-entry after months in space can be tiring even for a healthy astronaut, never mind a sick one, so Irene and her team of doctors worked hard to keep astronauts as healthy as possible.

Irene retired from her post just a few years ago. Throughout her long career, Irene has been committed to the well-being of everyone at NASA, keeping them all happy, healthy and achieving more than we ever could have dreamt.

'To know where you're going, you must know where you've been ... To succeed and prosper in the present, you must know where you're headed.'

LIVING

AND

WORKING

IN

SPACE

LIVING AND WORKING IN SPACE
2000 ⟶ NOW

On 31 October 2000, Bill Shepherd, Yuri Gidzenko and Sergei Krikalev launched into space and two days later arrived at the International Space Station, then just three modules joined together. This was the start of a permanent, unbroken human presence in space that continues up to this day. In 2003, China became the third country to put astronauts into space using their own rockets and spacecraft. Humans from all over the world have learnt to live and work together in space, performing hundreds of scientific experiments every year.

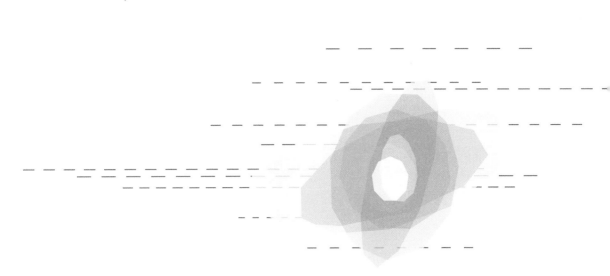

PEGGY WHITSON

⬢ BIOCHEMIST ◯ ASTRONAUT

⬤ USA BORN 1960 →

Peggy Whitson is a quietly confident introvert and one of the greatest astronauts in history. Throughout her life people told her it wasn't sensible to follow her dreams of flying to space – she ended up proving them very wrong.

Peggy grew up on a farm deep in the American countryside. When she was nine she was inspired by Neil Armstrong and the men who walked on the Moon. But it wasn't until she saw the first female astronauts that she thought she might give it a go too, so she sold chickens that she reared on the farm at a local market to save up the money to pay for flying lessons.

When she was at university, Peggy shared her dream of flying to space with a famous scientist and he told her that he thought astronauts weren't important and that it wasn't a good profession. But she never lost sight of her ambitions, and after years of study and working on experiments that flew on the Space Shuttle, her commitment and enthusiasm shone through and she was selected as an astronaut in 1996.

Peggy has flown in space three times, each time for approximately six months aboard the International Space Station (ISS). The ISS is an amazing feat of engineering, the biggest object ever constructed in space. It has about the same amount of living area for the astronauts as a five-bedroomed house, and is roughly the size of a football field.

FIRST FEMALE SPACE STATION COMMANDER

It took twelve years to construct, with each component part brought to the station by rocket or Space Shuttle. These were built all over the world, designed by different engineers, and joined together for the first time in space by a robotic arm. The different 'rooms' are called modules, and a huge 'spine' runs across it to hold the eight giant solar panels that power the station.

During her second stay on the ISS, Peggy was in charge, the first woman to be selected for the role of ISS Commander. It was a very busy mission, with three new modules arriving and two solar array panels moved to a new home. There was a big problem when one of the solar panels got torn during the move, but with Peggy's superb leadership skills and teamwork, the panel was repaired and the station assembly continued.

As well as being vital to the building and running of the ISS over the years, Peggy has completed a record ten spacewalks and in 2017 she broke the record for the longest cumulative time spent in space by any American astronaut, woman or man. Peggy is honoured by the records, but she is more proud of being part of the brilliant team of people that work to explore and learn about space.

'In terms of goals for NASA before I die, we need to be living on Mars. And I might not live that long, so they better get on with it!'

JULIE ROBINSON

SCIENTIST

USA BORN 1967 →

Julie Robinson thought she would spend her career working in universities, doing research and teaching students. She never expected to end up working in human space flight.

Julie is NASA's chief scientist for the International Space Station (ISS), representing the interests of thousands of scientists from over ninety-five countries worldwide who have experiments there, and working with partners in the other ISS space agencies. She studied chemistry and biology, and used her knowledge to understand what was happening in images of the Earth taken by satellites. This led her to work at NASA, looking at pictures taken by astronauts. Today, she makes sure that scientists can understand what is happening on Earth, by enabling them to do experiments off the Earth.

The ISS is a truly unique laboratory. Over 400 experiments are performed there every year, in numerous different scientific areas. Because the station – and everything inside it –

MANAGING THE SCIENCE THAT HAPPENS IN SPACE

is in permanent free fall, neither the astronauts nor the experiments can feel the effects of gravity. By studying how things behave in space, scientists can find out more about how materials react, how things work and what happens to the bodies of humans and other living things, making discoveries that could not be found out any other way. Scientists then apply all this knowledge learnt in space to things on Earth.

The astronauts carry out some of the experiments, but most are operated remotely by teams of scientists and engineers on the ground in mission control centres all around the world. Schoolchildren are also able to fly experiments to the ISS. One group, for example, suggested the astronauts try putting water next to something charged with static electricity, like a balloon that has been rubbed. On Earth, water will be attracted to the object and many people expected the same thing to happen. Everyone was surprised when the crew tried it, putting some drops of water near a statically charged rope, and the water ended up orbiting the rope.

Julie thinks this is a really exciting time for science in space. The ISS will stay in orbit until at least 2024 and possibly even later, so there is plenty of time for more research to be done. Scientists are making discoveries that challenge things we thought we understood, and we don't know what breakthroughs lie around the corner. Without the space station, and Julie making sure all the scientists are able to do their experiments, we would have a much poorer knowledge of things that will improve the lives of everyone here on Earth.

'It's up to us, the scientists, to continue to challenge ourselves to share this work with the public in new and dynamic ways.'

SUNI WILLIAMS

🪖 TEST PILOT 🧑‍🚀 ASTRONAUT

🇺🇸 USA BORN 1965 →

Suni Williams was really disappointed when she didn't get into her first choices for university. When her brother suggested that she go to the Naval Academy instead she wasn't sure – she was an Indian girl with long hair and didn't know what to expect. But having decided to give it a go, she discovered that there were plenty of opportunities ahead of her.

She loved flying helicopters, and became a test pilot. Through her training she found out what astronauts did, and realised it was something she could do as she had the right skills. She was thrilled to be selected as an astronaut in 1998.

During Suni's first mission, a six-and-a-half-month stay on the International Space Station (ISS) in 2006–7, she became the first person ever to run a marathon in space. She wanted to encourage children to make fitness an everyday part of their lives, so when the Boston Marathon was being run on Earth, she ran the equivalent 26.2 miles on the space station's treadmill. She had trained for the race just as everyone else did, by regular exercise.

Even if she hadn't been running the marathon Suni would have had to do lots of training. When people are in space, free-falling around the Earth and floating, they don't use their muscles as much or load their bones, and so these start to waste away as the body adapts to the new environment. If astronauts were to stay in space for ever, this wouldn't be a problem, but they need to be strong enough to get out of their spacecraft when they land on Earth. Usually there are recovery teams on hand to help, but if they had to come back in an emergency, this might not be the case.

FIRST PERSON TO RUN A MARATHON IN SPACE

To minimise the deterioration of bones and muscles, astronauts exercise for two hours every day: one hour of cardio activity and one hour of 'weights'. The ISS has a treadmill and a cycling machine, and a multi-use gym that uses resistance devices rather than normal weights. In the weightlessness of space, astronauts strap themselves to the treadmill using a harness that goes over their shoulders and is attached to the base by elastic bungee cords.

Suni has flown in space twice so far, breaking records and commanding the ISS. She's now busy helping to develop and learn to fly America's new spacecraft that will take crews to the ISS in the future.

'The types of people we look for when we select astronaut candidates [are] those who can challenge and push themselves, but who know when to stop before hurting themselves.'

JEANNE LEE CREWS

 AEROSPACE ENGINEER

USA BORN 1940 →

Jeanne Lee Crews has always been interested in how things work and loves solving puzzles. When she saw Sputnik orbiting the Earth in 1957, she knew she had to be a part of this exciting new venture and joined NASA as one of their first female engineers.

One day, a scientist named Burt Cour-Palais asked her to build a hypervelocity launcher, a special 'gun', to see how different materials could handle being hit by objects at very high speeds. Jeanne built the launcher and was fascinated when she saw what happened to the materials when they were hit. She and Burt worked together for the next twenty years to try and solve the problem of how to deal with debris in space.

Things in space move very fast. The International Space Station (ISS) orbits the Earth once every ninety minutes, travelling at 17,500 mph, about ten times faster than a bullet from a gun. If even a very small object were to collide with it, the damage could be huge.

Since the launch of Sputnik, over 7,500 satellites have been put into space. More than 4,000 are still in orbit, though only about 1,200 still work. Space is vast, and the chances of any of them meeting accidentally are rare; but there are millions of other objects in space, too. In the early days, people didn't worry about pieces of debris – used bits of rocket, springs off satellites – being left, but these small man-made objects can orbit the Earth for hundreds or thousands of years and now litter space.

PROTECTING SPACESHIPS FROM THE DANGERS OF SPACE JUNK

Jeanne and Burt designed a new debris shield, called a Stuffed Whipple Shield. This is used on the ISS to protect spacecraft from anything measuring 1 centimetre or smaller. Engineers track everything in space bigger than about 10 centimetres, so the ISS or satellites can move out of their way. But between 1 centimetre and 10 centimetres? As Jeanne said, 'You just have to hope and pray, and I don't much like that!' If the ISS were hit by one of these particles, it would make a hole. The air inside would likely take many hours to leak out, so the crew would have time to fix the damage or get in their spacecraft and come home, but it is something to be avoided.

As space gets busier and more crowded, the risks from debris will be ever greater. Jeanne's proposed giant 10-kilometre balloon to 'absorb' lots of small pieces was never made, but twenty years later, the need for such devices is getting more urgent. Her early recognition of the issue has made scientists work even harder to find a solution.

'I never did know what the words "You can't do it" meant in my whole life.'

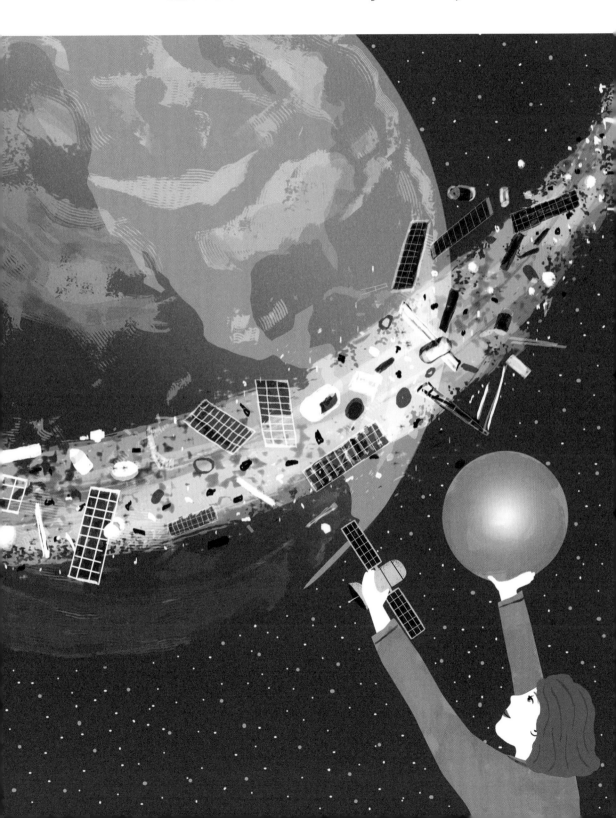

KALPANA CHAWLA

AEROSPACE ENGINEER · ASTRONAUT

INDIA · USA 1962 → 2003

LAUREL CLARK

DOCTOR · ASTRONAUT

USA 1961 → 2003

Kalpana Chawla grew up in India. She used to watch the planes at a local flying club, and hoped one day to become an aerospace engineer. She followed her dreams to the United States, where she got a job at NASA researching the behaviour of air flows on aircraft. Laurel Clark was a navy doctor. She had been trained for all sorts of challenging environments, carrying out medical evacuations from submarines, and working with aircraft as a flight surgeon.

Laurel and Kalpana were both selected by NASA as astronauts and were members of the crew of STS-107 on board the Space Shuttle Columbia. It was Laurel's first flight to space and Kalpana's second.

Columbia lifted off on 16 January 2003. Just eighty-two seconds after launch a piece of foam insulation separated from the external fuel tank and hit the wing. Teams on the ground spotted this, but similar events had happened before, so the programme managers believed the damage was minor and would not cause Columbia any problems, and reassured the crew that all was in order.

THE COLUMBIA ACCIDENT

Kalpana, Laurel and their five crew mates, Rick Husband, Michael Anderson, William McCool, Ilan Ramon and David Brown, had a really busy mission. They spent sixteen days in orbit, working in shifts to carry out experiments twenty-four hours a day. Among them, Laurel observed roses bloom and silkworms hatch. Kalpana helped fix an experiment looking at how water mist might stop a fire in microgravity.

On 1 February, Columbia fired its main engines and headed home after a successful mission. As the Shuttle re-entered the Earth's atmosphere, hot gases rushed into a hole in the wing where the foam had hit on launch. Mission control started seeing some strange readings in their data, and then lost communication with the Shuttle. A few minutes later they heard reports of multiple streaks of light over Texas, where Columbia should have been flying through the atmosphere. Columbia had been torn apart by the hot gases rushing into the wing, ultimately causing a catastrophic loss of control and the disintegration of the vehicle, killing all the crew.

Once more NASA and the space community had to confront the dangers of space flight. All of the Shuttles were grounded while the accident was investigated. NASA decided to retire the Space Shuttle once it had finished building the International Space Station and then focus on developing safer spacecraft. Laurel, Kalpana and the crew's mission ended in tragedy, but it was not in vain. Their experiments led to advancements in many areas of science, including helping to improve cancer drugs, and their research will live on, keeping their memory alive.

'When you look at the stars and the galaxy, you feel that you are not just from any particular piece of land, but from the solar system.' KALPANA CHAWLA

'There was a moth [in one of the silkworm cocoons] ... just starting to pump its wings up. Life continues in lots of places and life is a magical thing.' LAUREL CLARK

NICOLE BUCKLEY

SCIENTIST

CANADA BORN 1960 →

Nicole Buckley has always been curious about the world around her. When she was young she thought all sorts of jobs might be fun, from actor to firefighter, spy to nurse. One day she read H. G. Wells's novel *The War of the Worlds*, about Martians coming to take over the Earth. They are defeated when they finally venture out of their spaceships and their immune systems cannot handle our bacteria. This inspired Nicole to learn more about biology.

She studied microbiology and as she was completing a research post, started looking for jobs. The Canadian Space Agency (CSA) wanted to hire a scientist. Nicole liked the sound of the job but thought, 'I don't know anything about space.' When a friend said that probably not many people did, and that nobody knew if there was life on Mars, Nicole decided to apply. She was thrilled to be accepted.

Nicole is now the chief scientist for all life sciences at CSA, responsible for planning the future of life sciences research, deciding what should be studied and how. She hopes that one day, as engineers invent new technologies to give the world cheaper and more reliable access to space, all scientists will look at space as just another place where research can be done.

Nicole's first mission as programme scientist was Space Shuttle Columbia's STS-107, which turned out to be Columbia's last. A hole in the heat shield let hot gases into the craft during re-entry, causing the Shuttle to disintegrate and her crew all perished. Nicole found herself asking if the science being done was worth the dangers of space travel. She believes that no one's lives should be lost in the pursuit of knowledge so she resolved to make sure that only the very best experiments in space would make it to space, in honour of the risks taken by the crew.

Canadian experiments have made many great contributions to science, including in the field of medicine. CSA put devices inside astronauts' spacesuits to monitor how much radiation they were exposed to. This technology is now being used to determine how much radiotherapy cancer patients should receive and to make sure it is directed to exactly the right place, thus improving treatments and limiting side

SPACE RESEARCH FOR THE BENEFIT OF ALL MANKIND IN SPACE AND HERE ON EARTH

effects. The technology developed for Canadarm, the amazing robotic arm built for the Space Shuttle and the ISS, is now used in the robotics with which surgeons operate on brain tumours. Nicole makes sure that the research Canada does in space helps humanity here on Earth.

'The world is so much vaster than you can imagine. Keep your eyes peeled, your options open.'

BERTI BRIGITTE MEISINGER

⚙ ENGINEER 📊 MISSION DIRECTOR

● GERMANY 1958 →

When Berti Meisinger was a young girl, she dreamt of working on a ship and sailing the high seas. As women weren't allowed to be captains at that time she wanted to be a communications officer and work closely with the captain to sail the ship, so that she could be at the centre of all the action.

Near her home in the countryside outside of Munich, Germany, was a big campus with lots of buildings. As she grew older, she found out that this was a space institute, and decided that working with spaceships would be much more interesting than working with sailing ships. Her determination earned her a placement at the German Space Agency, called the Deutsches Zentrum für Luft- und Raumfahrt (DLR), when she was still at school and she has remained on the campus ever since.

Berti started out working with the ground segment, which is all the equipment that allows mission control centres to manage satellites and receive the data collected by them. Within a few years Berti's expertise meant she was in charge of it all – antennae, radios, computers – and her bosses sent her all over the world to train other engineers in how to use the equipment.

Next, she started learning how to fly spacecraft, including working on the experiments that went on some Shuttle missions. In 2000, Berti was in charge of the German part of an international experiment that flew on the Space Shuttle and produced the first ever global 3D radar map of the Earth. She made sure the mission was planned very carefully, down to the minute. At one point, as the Shuttle was about to fly over Germany, disaster struck and the recorders stopped, but thanks to Berti's advance planning, mission control knew that switching them off and on again would make them work. The mission continued successfully.

READY FOR ANYTHING, IN CHARGE OF THE MISSION

Berti is very knowledgeable and a fantastic planner and go-getter. She was the perfect choice as one of Europe's first mission directors, in charge of operating the European part of the International Space Station, the scientific laboratory called Columbus. The mission directors have to keep all the different teams who control Columbus working together, to make sure that as much as possible of the planned mission happens. Berti thinks being a mission director is like watching a thriller movie – you never quite know what will happen. You must prepare for the worst but expect the best. She always makes sure that her teams do all they can to keep Columbus flying efficiently and deliver the best every time.

SAMANTHA CRISTOFORETTI

🛡 FIGHTER PILOT ⚒ ENGINEER 🧑‍🚀 ASTRONAUT

🇮🇹 ITALY BORN 1977 →

Samantha Cristoforetti was prepared in astronaut training for all the eventualities that space travel can bring. So when a cargo craft had a problem delivering supplies and she was asked to stay in space for longer, she didn't bat an eyelid and got to work. In the process she became a record-breaker.

Samantha always knew she wanted to go into space. She grew up in a tiny village in the Italian mountains, and remembers how seeing brilliant bright stars in the night sky helped spark her interest in the cosmos. However, the European Space Agency (ESA) doesn't recruit astronauts very often. She decided to follow her other goal of flying, and joined the Italian Air Force, going on to be a trailblazer as one of Italy's first female fighter pilots. She had just completed her training when ESA spotted her talent and recruited her as an astronaut.

Samantha was launched into space in November 2014, along with crew mates Anton Shkaplerov and Terry Virts. Each group of people working on the International Space Station (ISS) is given a mission number, called an Expedition, and Samantha's crew was called Expedition 42. The number 42 is famously declared to be the answer to 'the great question of Life, the Universe and Everything' in Douglas Adams' novel *The Hitchhiker's Guide to the Galaxy*, which also says that a towel is the most useful thing for an interstellar hitchhiker. Samantha and her crew paid homage to the book throughout their mission, and celebrated Towel Day on 25 May.

HOW LONG CAN YOU SPEND IN SPACE?

Their five-month mission on the ISS went according to plan, with the crew helping to run hundreds of science experiments and even installing a coffee machine. They were just days away from heading home and making final preparations for a return to Earth when there was a problem. A cargo ship which was bringing supplies hadn't made it into space as planned. Mission control did not want to launch the next crew until they understood what the problem was so Samantha and her crew mates were asked to stay for longer. This meant that she spent 199 days and 16 hours in space, the longest flight by any European astronaut.

It takes about a year for a human being to recover from being in space for six months. On her return, Samantha said her legs felt as heavy as tree trunks. Her long stay in space meant that not only could important experiments continue while she was there, but she could also help scientists understand how being in space affects the human body.

'When you discover new things every minute
your mind is absorbing so many experiences,
it feels like time expands.'

NORIKO SHIRAISHI

 ENGINEER

JAPAN BORN 1976 →

Noriko has been guided through her life by following her dreams and taking on the biggest challenges. As a child she looked up to the sky and wanted to be able to fly, so she went to university in Tokyo and studied aircraft. Whilst there, she learnt about how difficult it was to get a rocket to fly, and thought what a brilliant challenge it would be to work on.

Noriko was determined to follow her rocket dreams so she joined the Japan Aerospace Exploration Agency (JAXA). Here she saw a launch conductor at work and aspired to being that 'because it was a tough job, but worthwhile'. Seven and a half years of intensive training later, she became the first female launch conductor.

Noriko was in charge of the team responsible for launching the H-IIB rockets. These launch the Japanese cargo vehicles called HTV, carrying vital supplies and equipment to the International Space Station (ISS). She and her team managed every launch from a bunker just 500 metres away from the launch pad, but 12 metres underground so they are safe if the rocket explodes.

The ISS supports six crew members, who usually stay for six months each. Everything they need when there – food, clothes, toilet roll, experiments to work on, personal items – must be flown into the ISS in cargo ships, such as HTV. Some water is delivered, but most is recycled on board the station, treating the astronauts' urine as well as condensate from the air-conditioning units to turn it back into drinking water. When the HTV has delivered its cargo, the crew fill it with their rubbish, dirty clothes, broken equipment and even all the bags of dried human waste. This spacecraft has no heat shield, so it burns up with all its contents on re-entry to the Earth's atmosphere.

As a launch conductor, Noriko gets to push the button which launches the rocket; she also has to push the emergency stop button if something goes wrong. This is a nerve-racking responsibility and Noriko has to be able to make quick decisions, with complete trust in her team of about 150 people. After the launch she evaluates all the data, looking for any problems and making sure the next flight is even better.

LAUNCHING SUPPLIES TO THE INTERNATIONAL SPACE STATION

Noriko hopes one day to build a Japanese rocket that can carry people. She's currently working on Japan's next rocket, the H-III, which will be bigger and better, a fantastic home for her drive for excellence. Noriko will keep following her dreams and perhaps one day she'll be launching humans into space.

'Focus on what you are interested in and I
believe you will be able to pursue your dreams.'

ANOUSHEH ANSARI

↓⚙ ENGINEER 🧳 SPACE TOURIST

🇮🇷 IRAN 🇺🇸 USA BORN 1966 →

Anousheh Ansari grew up devouring science fiction novels and *Star Trek*. When she said she wanted to be an astronaut, no one took her seriously. Her home country, Iran, had no space programme or agency, and everyone assumed she would grow out of the dream. Instead, her vision kick-started a space revolution.

When she was sixteen, her family moved to America to escape the unstable political situation in Iran. She went to university, studied electronics and computer engineering, and met her husband. They set up a company that created a hugely successful product allowing voice communications over the internet, and made hundreds of millions of dollars.

Anousheh had never forgotten her dream to go to space, or her belief that one day Starfleet Academy would exist for real, so she decided to use some of her fortune to fund a prize that might change the way people think about space travel.

The Ansari X Prize offered $10 million to the first non-governmental organisation to fly the same crewed spacecraft 100 kilometres into space and back, twice within two weeks. Twenty-six different groups entered the competition, some backed by

FIRING UP YOUNG WOMEN TO FOLLOW THEIR DREAMS

big companies, others much smaller groups working in their spare time. The prize was won in October 2004 when SpaceShipOne made two flights just five days apart, reaching altitudes of 103 and 112 kilometres. The winning spacecraft was then bought by Virgin Galactic, who wanted to develop the technology and offer private citizens the opportunity to head into space. As soon as tickets went on sale, Anousheh bought hers, her dream of going into space now a future possibility.

A few years later, though, another avenue to space became possible when Russia started selling seats for a short trip to the International Space Station. Anousheh could afford the astronomical cost, and arranged to be backup to a Japanese businessman, Enomoto Daisuke. She spent six months training in Russia, just in case. Only a few weeks before Enomoto was due to blast off Anousheh got a call – he could no longer fly due to medical issues. She couldn't believe it and wanted to scream with excitement.

Anousheh lifted off on 18 September 2006 and spent eight days visiting the ISS. She performed experiments for the European Space Agency and shared her experiences via her blog. Since her return from space, she has gone back to her business but also works to encourage more young women to follow careers in science and technology, and to make childhood fantasies come true.

'You can make the future whatever you want to do with it, whatever you want to be ... Everything around you is there to empower you to shape that future.'

GINGER KERRICK

 FLIGHT DIRECTOR

USA BORN 1969 →

Ginger Kerrick vividly remembers the day when, aged five, she got a book about astronomy and space from the library. She can still see its cover, and the dream it sparked to be an astronaut.

At thirteen, Ginger wrote to NASA, saying, 'I want to work here.' NASA replied with the advice to stay in school, work hard, study science and maths, and participate in team activities. Ginger followed this to the letter and it paid off. Her first job at NASA was teaching astronauts about the systems used on the International Space Station but she still longed to be an astronaut herself. As soon as she was able to, she applied for selection; from 3,000 applications just 120 people were invited for interview, an excited Ginger among them.

That high soon turned to devastation. During the medical part of the selection process, doctors found out that she had kidney stones, meaning she could never be an astronaut. Ginger went home miserable. How could she go back and teach astronauts knowing she would never join them? But then she had a stern word with herself: 'Get over it, get a grip.' She realised that by teaching astronauts, a part of her would still be going into space.

Ginger took every opportunity that came her way. As NASA prepared for the launch of the space station, they wanted two people to train alongside the astronauts, and she convinced her bosses she should be one of them. For four years she went through astronaut training, leading to even more exciting possibilities.

The voice of mission control to the astronauts in space is called CapCom. Just one person talks to those in space, and since the beginning of mission control, this had only ever been done by other astronauts. But with all her training, Ginger was the perfect fit and she was appointed to the role – the first non-astronaut ever.

CapCom sits next to the flight director, who is in charge of the large team of people at mission control and makes all the decisions, protecting the crew, the spacecraft and the mission. They plan the operations for every day, decide what should happen if anything goes wrong and have backup plan after backup plan. Watching from her CapCom seat, Ginger desperately wanted the job – and this time she got it.

LEADING THE TEAM IN MISSION CONTROL, HOUSTON

Ginger works alongside the astronauts every day. Over 800 people have flown in space, but fewer than 100 have sat in the flight director's chair in Houston's mission control. Ginger's life didn't take the course she had hoped, but she is very happy about that and how things turned out.

'There are going to be bumps in the road ... You're going to need to divert your course ... And that's OK ... because you might be very pleasantly surprised at what lies in store for you on that new path.'

ELENA SEROVA

AEROSPACE ENGINEER   COSMONAUT   POLITICIAN

RUSSIA BORN 1976 →

Elena Serova flew into space on 26 September 2014, the 540th person to make it there. Most of those represent the USA and Russia, but over thirty other countries have also had people fly their flag in space. With all the milestones that Russia have achieved in the history of human space flight, you might think lots of the women in space would be Russian. But when Elena headed to the International Space Station, she was just the fourth Russian woman in space and the very first to visit the ISS.

Elena was born in a little village in Russia. Her dad used to take her to watch planes soaring through the air, inspiring her to set her sights on the stars. She felt so strongly about reaching the glittering night sky that she would often dream she was flying. She went on to specialise in test engineering at an aviation school, and then had jobs both at a spaceship maker and at mission control. Her skill and passion saw her selected as a cosmonaut and as she blasted off on that September day, she was finally living her dream.

FIRST RUSSIAN WOMAN TO VISIT THE INTERNATIONAL SPACE STATION

Traditionally, Russian culture sees cosmonauts as a typically male role, but Elena thinks gender makes no difference. She is not a 'woman cosmonaut', just a cosmonaut, but the media became fixated on her gender, asking lots of questions about how, as a woman, she would cope. When in an interview with her male crew mates just before launch she was asked how she would manage her hair, she fired back with her own question: 'Aren't you interested in the hairstyles of my colleagues?'

Elena proved her talents by going on to do her job superbly. She stayed on the ISS for 170 days, working on many experiments, observing the Earth and maintaining the station. She also decided to carry out her own experiment. When cargo ships deliver new supplies to the ISS, ground crew put fruit and vegetables in so the cosmonauts can enjoy some fresh food; on Elena's flight this included apples, which she loves. She decided to see if the seeds would grow in space: they did and she nurtured a little seedling for several weeks.

Elena retired from the cosmonaut corps in 2016 when she was elected to serve in the Russian Parliament. Yet again she has proved that women can work in whatever job they wish and dreams really can become a reality.

'We do it for others, for future generations, so that people on Earth live better, have better opportunities.'

THE FUTURE OF SPACE
NOW →

Space is changing. Private companies are developing new vehicles to take fare-paying passengers on out-of-this-world journeys. The International Space Station is getting older and will eventually be retired and perhaps replaced, and China are planning their own outpost in the coming years. Space agencies and others are working on plans to send humans away from the relative safety and security of Earth orbit, perhaps back to the Moon and then on to Mars, searching for evidence of life on the Red Planet and proof that we are not alone in the universe.

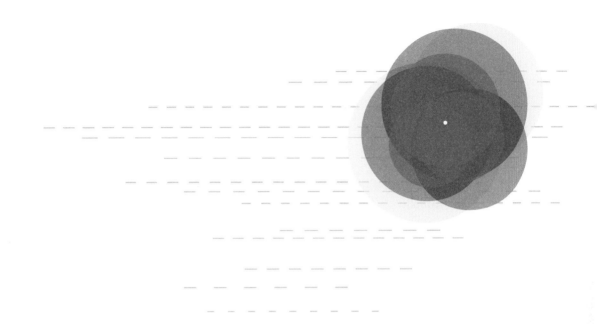

LIU YANG

🛰 PILOT 🧑‍🚀 TAIKONAUT

🌏 CHINA BORN 1978 →

WANG YAPING

🛰 PILOT 🧑‍🚀 TAIKONAUT

🌏 CHINA BORN 1980 →

Liu Yang and Wang Yaping, the first female taikonauts, or Chinese astronauts, are leading the way in a new chapter of space exploration. Since the flights of Yuri Gagarin and Alan Shepard in 1961, hundreds of people have flown in space, but until 2003 all had been in American and Russian rockets and spacecraft. China then joined them with the flight of Shenzhou 5.

Liu and Wang were both military pilots who were selected in China's second group of taikonauts. The selection process was incredibly rigorous, demanding superhuman-like levels of perfection, and not only fitness, courage and the ability to work well in a team, but also no physical scars, snoring or strong accents.

After they and the rest of their class were selected, the strict training was conducted in great secrecy. Their names were not announced and only their parents knew they were planning to go to space. Some brilliant detective work was carried out by people outside China when a signed postcard was found for sale – a commemorative item that had accidentally been released early, and revealed the seven names of those in training.

FIRST CHINESE WOMEN IN SPACE

A few months before Shenzhou 9 was due to blast off, China announced that the three-person crew would include China's first woman in space, but did not say whether it would be Wang or Liu. It was only on the day before launch, 16 June 2012, that they confirmed to the world it would be Liu Yang. She and her crew mates, Jing Haipeng and Liu Wang, were the first people to visit China's small space station, Tiangong-1. They stayed there for eight days, performing experiments and testing out the new laboratory.

Wang Yaping flew to Tiangong-1 a year later. During her fifteen days there, she delivered China's first live science lesson from space, broadcasting to 60 million children across the country. She showed how different objects behave in the weightless environment, and confirmed that she hadn't seen any UFOs.

China have big ambitions for their growing space programme, having already put a second small station, Tiangong-2, in orbit and flown a cargo ship to it. There are plans for a bigger space station, to send a robot to the Moon to bring back some rocks, and one day to send taikonauts to the Moon and Mars. It is not known if Liu or Wang will journey into space again but they have shown that women can play a vital part in China's missions.

'I believe in persevering. If you persevere, success lies ahead of you.'

LIU YANG

'When I looked out of the window for the first time, I realised the true meaning of the power of life ... That kind of beauty was just beyond comprehension.'

WANG YAPING

KELLY LATIMER

TEST PILOT

USA BORN 1964 →

Kelly Latimer never makes a flight without being prepared and having a plan. When her life plan went off course she didn't give up and the outcome was unexpectedly positive.

TESTING A NEW GENERATION OF COMMERCIAL SPACESHIPS

For as long as she can remember, Kelly wanted to be an astronaut. When she was at school she did lots of research and found out that most astronauts have been military test pilots so she set her sights on attending the military academy. Having only been a passenger on a commercial flight a couple of times, she assumed her parents would think she was crazy, so applied in secret. Only when she was called for a medical and needed their help to get there did she tell them. Happily, they were really excited and supportive.

Kelly was successful and in 1983 headed to US Air Force Academy. When at the end of the first year she started learning to fly, she worried and thought, 'I really hope I like this – I've planned my whole career on it.' As soon as she took the controls on her first solo flight she knew she had made the right choice.

After graduation she headed to the US Air Force Test Pilot School, and was one of the first handful of women to become a test pilot. As soon as she could, she applied to be an astronaut. She made it through to the last thirty-five candidates, but in the end wasn't hired. Naturally, she was upset but she didn't give up. Four years later, Kelly applied again, but when she failed to pass the medical tests for an extended stay in space, she knew her dream was over. She had to accept that she would never be an astronaut.

Years later, Virgin Galactic were looking to hire test pilots to fly their first commercial spacecraft, SpaceShipTwo, which will take fare-paying passengers on a sub-orbital rocket trip into space and back. Out of the blue, Kelly got a phone call. This time she passed the medical tests and now her dream of going to space is alive again, and even better. She is currently a test pilot for the carrier aircraft which will take SpaceShipTwo high up into the atmosphere and in the coming years will qualify to fly SpaceShipTwo as well.

Not only will Kelly go into space herself, she will also get to take other people up who, just like her, have dreamt of space. She cannot wait to feel the acceleration when the rocket fires, and then the sudden cut-off and weightlessness, looking back at Earth. Kelly's plan is working out after all.

'Do the best you can at whatever job you are doing. Anything you do will give you experience.'

GWYNNE SHOTWELL

⚙ ENGINEER 📈 COMPANY PRESIDENT

🇺🇸 USA BORN 1963 →

At school Gwynne Shotwell loved maths and science and finding out how things worked. She thought about a career in engineering but was afraid that she would be called a nerd as it wasn't cool. One day her mother took her to an event where she met a mechanical engineer who owned her own business and Gwynne saw that she was a brilliant person. She realised that being a female engineer wasn't at all a strange thing to do, and she has spent her career doing some very cool things.

Gwynne has worked in the space business ever since she left university, starting off as an engineer and then taking on management roles. In 2002 she joined SpaceX, a young company with big goals to develop new rockets and spacecraft, when they were still tiny and just six years later became their president.

PLANNING TO GET HUMANS BACK TO THE MOON AND ON TO MARS

Gwynne's role is to be responsible for all the day-to-day operations, using her engineering skills as a manager. SpaceX are developing new rockets and spacecraft which will soon be ferrying crew to the International Space Station, but they are planning on going further. They have announced plans to send humans to lunar orbit and hope one day to send humans to Mars.

As SpaceX have developed new rockets, they have had to invent all sorts of new technologies; they knew this would not be easy, but they weren't afraid to give it a go. They have tested things over and over again, even when they suspect that they will not succeed. Sometimes their tests do end in failures, but they learn a lot from these and improve their designs as a result. As leader of SpaceX, Gwynne constantly pushes her team forward. Even when their new technology for landing rocket stages had had five good tests, Gwynne still said, 'Come on, we're not pushing hard enough,' knowing that perhaps they were playing it safe and that there was still more to learn.

SpaceX are revolutionising the world of space travel. They have managed to fly the first stages of their rocket back to Earth and carry out a pinpoint landing on a target on a barge in the middle of the ocean, a bit like trying to stand a pencil on its flat end. This is driving down the cost of rocket launches, which means that space travel should get cheaper and more accessible to all. Gwynne is pushing SpaceX hard so that in the not-too-distant future you will be able to buy a ticket and go into space yourself.

'Do great work. I believe that you get recognised when you do great work. Figure out what you are there to do and do a great job.'

FRANCES WESTALL

 GEOLOGIST

SOUTH AFRICA UNITED KINGDOM FRANCE BORN 1955 →

Frances Westall is a geologist who has spent her life studying the oldest known rocks on Earth, looking at the evidence of tiny beings that once lived in them. She has also been investigating the possibility of tiny beings in even more unexpected places.

Ever since people first thought about going into space, one destination has always held a great fascination: Mars. It is similar to Earth in some respects, nearby, and for centuries we have wondered if there is life there. Spacecraft first started visiting Mars in the 1960s and though we know there are no little green men, scientists still think there could have been primitive life forms there, and maybe there still are somewhere under the surface.

Currently, the Earth is the only place we know of in the whole universe that supports life, but we also know that the universe is unimaginably large. Our Sun is one of about 100 billion stars in our galaxy, and astronomers estimate that there are at least 100 billion galaxies in the universe. Most scientists think this means that somewhere out there, at some point, life has evolved in one form or another. Finding evidence of that will change the way we think of our beautiful blue and green planet – and one of the places we can look is Mars.

IS THERE LIFE ON MARS?

In the beginning the Red Planet was much more similar to Earth than it is today, but scientists think that about 3.5 billion years ago the surface of Mars became inhospitable. Life might, however, have survived underground by extracting nutrients from the rocks, just as some creatures on Earth do today. Frances has studied fossilised bacteria, the oldest known form of life on Earth. She has done lots of experiments to show how these life forms can be preserved in their underground habitat, and has looked at rocks to find out what early life was like and might be like on Mars.

Frances is one of the scientists working on the European Space Agency's ExoMars rover, due to head to the Red Planet in 2020. It will use a drill to search for signs of life. She will be analysing the data very carefully – but to really use her skills she would like to study the rocks herself. She is looking forward to the day when samples of Martian rock can be brought back by robots or even humans, and maybe she will find evidence that life has evolved somewhere other than Earth.

'If life ever did appear on Mars, studies of its evolutionary steps will greatly help us understand the early evolution of life on Earth.'

SIMONETTA di PIPPO

ASTROPHYSICIST

ITALY BORN 1959 →

Simonetta di Pippo's passion has always been space. She remembers watching the Moon landings when she was young, and found humankind's ability to escape the limits of Earth very inspiring. Throughout her life she has been on the lookout for jobs that enable her to keep her passion alive.

Simonetta studied astrophysics at university and found herself working for the newly formed Italian Space Agency (Agenzia Spaziale Italiana, ASI). She has spent her career in the international aspects of space, working with different countries, including being in charge of human space flight for the ASI and then the European Space Agency (ESA). All of this prepared her well for her current post as director of the United Nations Office for Outer Space Affairs (UNOOSA).

UNOOSA encourages better international cooperation in space activities. Simonetta works tirelessly to help countries see how space can benefit humanity. By working together and using space for peaceful purposes, we can tackle some of today's challenges such as climate change, energy creation and natural resources management.

UNOOSA also works to ensure that countries uphold the five international space treaties which set out the peaceful use of space. As different nationalities start making plans to explore the solar system again, these will be very important. One of the articles states that any exploration done by humans must avoid harmful contamination, so we do not introduce any diseases that could ruin a planet's natural habitat.

Scientists agree that we should preserve the record of history that exists on other planets and not introduce life from Earth that might destroy or weaken any potential life there. Agencies around the world have worked together to write rules about making spacecraft free

PROTECTING PLANETS FROM ALIEN LIFE

from anything potentially harmful. Any craft landing on Mars must be sterilised to remove microbes, and if it is going to a region where there might have been life the standards are even more stringent. However, if humans were to visit Mars, it could be extremely difficult to avoid contaminating the planet with Earth life, which raises the question of whether we should even send humans there, especially if we are trying to find signs of life.

Simonetta's work keeps planets protected and space peaceful; it even tackles issues here on Earth. She also strives to help other women find their place in space science. She set up and is president of an organisation called Women in Aerospace Europe, devoted to improving representation of women in aerospace and helping the youngest find their own career paths.

124

'I believe that true passion has no gender and focussing on your goals and dreams helps you to overcome gender and cultural boundaries.'

ELLEN STOFAN

GEOLOGIST PLANETARY SCIENTIST

USA BORN 1961 →

Ellen Stofan went to see her first rocket launch aged just four. She watched as it exploded on the launch pad in a terrifying blast of power and flame, which, she says, is probably why she never wanted to be an astronaut. Instead she wanted to be an archaeologist, but one day her mother took her to a course she was studying on geology. Ellen was hooked, picking up rocks and looking to see what they could tell her. As she got older, she learnt that you could study rocks from other planets. She knew that was what she wanted to do.

HOW DO WE GET TO MARS?

Ellen has had an amazing career, focusing on the geology of Venus, Saturn's moon Titan, Earth and Mars. By studying the other planets in our solar system, scientists can understand more about how the Earth works; the ozone hole was first found by scientists mostly studying Venus. Ellen was appointed as NASA's chief scientist in 2013, so she really knows her stuff, and during her time in the post she was key to developing long-term plans for getting humans to Mars.

This international task is being planned now, with the first steps already in place. Astronauts on the International Space Station are testing technology such as how to recycle urine into water and how to grow plants there. Recent Mars rovers have collected data about radiation, which is critical to understand if we are going to send humans there.

WHY SEND HUMANS TO MARS?

Worldwide space agencies are working together on plans to put a small space station into orbit between the Earth and the Moon, where the science and technology needed for sending humans to Mars can be explored and tested. NASA and the European Space Agency are building the Orion spacecraft, which will one day be able to take humans on long missions. Together, the agencies plan to send humans to Mars in the 2030s.

Why send humans at all? We've sent rovers in the past and continue to do so; they are cheap and easy to fly. But Ellen is convinced we need to send scientists to Mars. Humans can read a landscape in detail, observe and reason, and work much more quickly than robots. They can select the right rocks to bring back to a lab, choose which investigations to do there and try to find the answers. Only by sending humans, Ellen believes, will we be able to definitively answer the question of whether there is, or has been, life on Mars.

'When you push ... that's when you get
your leaps forward.'

MONICA GRADY

?¿ SPACE SCIENTIST

✠ UNITED KINGDOM BORN 1958 →

Monica Grady grew up in Leeds and spent her weekends in the Yorkshire Dales, looking at the part rocks played in the history of its beautiful landscape.

Her interest in rocks led her to study chemistry and geology, as part of which she examined some of the Apollo Moon rocks – 2,415 samples brought back from the six Apollo missions that landed on the Moon between 1969 and 1972.

Monica thinks that the Moon rocks are extremely beautiful. When geologists study rocks, they take a thin slice and polish it, so that under a microscope they can see what it is made from. Those from Earth are cracked, broken and rusty coloured, where they have been battered by the wind and rain. The Moon is a very different place, with no atmosphere and no weather, and the surface has remained the same for most of its 4.5-billion-year life. The only changes are the craters formed by rocks that hit the Moon after hurtling through space. When Monica looked at thin slices of Moon rock she saw that they were deep and clear, full of bright colours – cerise, turquoise, pink, yellow, grey – with sharp outlines telling of their origins.

Monica has become a leading expert in meteorites, particularly those from Mars. Rocks travelling through space rain down on the Earth all the time, mostly from the asteroid belt. Small ones burn up in the atmosphere, which we see as shooting stars. Bigger ones which make it all the way to the surface of the Earth are called meteorites. Every few million years a large object, several kilometres in diameter, crashes into Mars, forming a huge crater and scattering rocks. Some of these are flung out into space, and slowly, many millions of years later, they hit the Earth, having survived the fiery plunge through the atmosphere. Monica is enraptured by these – of a meteorite she says, 'If I break it open, it's thousands of million years old; no one else has ever seen it and it tells you about the history of the solar system.'

MOON ROCKS AND METEORITES FROM MARS

For years Monica curated the UK's collection of meteorites at the Natural History Museum, and is now working to make a place to store rock samples that will be brought back to Earth by future robots exploring the Moon, Mars or asteroids.

Monica has had an asteroid named after her in recognition of her brilliant work. Luckily for us though, it's in a stable part of the asteroid belt and is very unlikely to become a meteorite and come anywhere near Earth.

'When you are working on a project, every team member is so important ... Being able to work in a team where you have a common goal, now that is just really exciting and rewarding.'

GERDA HORNECK

SCIENTIST   ASTROBIOLOGIST

GERMANY   BORN 1939 →

Gerda Horneck has always been at the forefront of scientific discovery, looking at science that others thought was just the stuff of fantasy. As technology develops we get closer to making fantasies into realities, to reaching planets we thought might be beyond our reach, and Gerda has helped to drive this big leap forwards.

When she was a student, the idea that perhaps the first tiny organisms had come to Earth on a meteorite from another planet was difficult to believe. But the possibility of little hitch-hikers from faraway planets fascinated her. Gerda has spent much of her career looking at how all forms of life cope with being in space.

DID LIFE ORIGINATE IN SPACE?

Life on Earth is protected from harsh cosmic and solar radiation by the planet's magnetic field. This deflects harmful particles that come streaming towards us from the Sun and other sources outside the solar system. But beyond the reaches of this protective bubble, several thousand kilometres above the surface of the Earth, there are much higher levels of radiation that can be lethal to humans.

For future astronauts travelling to Mars, this is a problem. With current technology the journey takes about nine months and once you reach Mars, you have to stay for about six months before turning for home, while the planets line up again in their orbits around the Sun. Mars doesn't have a magnetic field like the Earth, and though the atmosphere provides some shielding, the radiation levels are about 100 times greater than those on Earth. Scientists and engineers around the world are working on inventing materials to protect astronauts from these harmful rays.

Many people thought it was impossible for any living thing to survive in the harsh environment of space. Gerda was one of the pioneers in measuring the heat and radiation and their effects on living organisms. One of her experiments put samples of bacteria, fungi and ferns in a container on the outside of the International Space Station for almost two years, to see what happened and whether these organisms might survive on a meteorite travelling through space. Some did.

Gerda has contributed so much to her field that she has had a bacteria named after her, *Bacillus horneckiae*, in honour of all her achievements. Her work has helped people understand more about where life might have come from, and will help keep astronauts safe when they venture out into the solar system.

'Astrobiology has become very popular, and I think maybe I have contributed a little bit to that.'

ANITA SENGUPTA

 AEROSPACE ENGINEER

USA BORN 1977 →

Anita Sengupta has been surrounded by engineering her whole life. When growing up she enjoyed helping her dad fix things around the house. One day, when the house had lost water pressure, after hunting for the source of the problem Anita found a burst water pipe. Her dad said, 'You'll be a great engineer one day!' He was right.

Throughout her career, Anita has not been afraid to take risks and has grasped every opportunity, even if it hasn't been what she was expecting or has been something she knew little about. Her first projects involved working on rocket engines. She ran computer simulations of how the gases and liquids would behave, including work on Space Shuttle fuel tanks.

One day, she was approached by the team developing Curiosity, a rover that would land on Mars. They needed help working out the parachute design and thought Anita would be just the person. It would involve dropping out of helicopters into the desert to test the parachutes. She didn't know anything about parachutes, but figured she had the right skills and could learn on the job – plus it sounded like a lot of fun.

The Martian atmosphere is 100 times thinner than Earth's and the parachutes would open at supersonic speeds. All those that had been used on Mars previously had failed to work properly, but no one had figured out why. Anita and the team came up with a new design, tested it over and over again, and when Curiosity landed on Mars, the parachutes worked perfectly.

LANDING SAFELY ON OTHER WORLDS

When NASA started developing the Orion, the next generation of spacecraft that will fly humans to the Moon and Mars, Anita could see they were going to have similar issues, as the re-entry speeds were so high. She got in touch and offered her expertise. The programme managers were delighted to have her knowledge and Anita came on board.

On Orion's first test flight on 5 December 2014, all the key design elements, including the parachutes, worked perfectly. They slowed the spacecraft to splash down safely in the Pacific Ocean.

Anita's latest project is leading the design, development and operations of an experiment that will fly to the International Space Station. The Cold Atom Laboratory is designed to study very cold quantum gases, testing some of the most fundamental laws of physics and observing phenomena not possible on Earth. She still seizes every opportunity, solving problems to push the limits of space exploration.

'There's a job for you even though it's a really tiny field. If this is what you want to do, and as long as you're motivated and trained, you'll get it.'

THE FIRST PERSON ON MARS

The first person to set foot on Mars is probably alive right now. That person might be reading this book. That person could be you.

Space agencies and private companies around the world are building spacecraft capable of sending humans to Mars and working on ideas for how to land them there. By the 2030s we may see humans in orbit around the Red Planet, with landings to follow in the years ahead.

There are still, however, a lot of challenges and issues to overcome.

With current propulsion technology, a journey to Mars takes about nine months each way. The crew will need shielding from the harmful cosmic radiation that will bombard them throughout the journey. When they reach Mars their landing craft, and other cargo vehicles bringing supplies and equipment, will have to slow down and descend through Mars's thin atmosphere to land safely on the surface.

The crew will have to survive for six months, while they wait for Earth and Mars to align in their orbits again before they can leave. They won't be able to get back in an emergency so they will have to be entirely self-sufficient, able to respond to anything that goes wrong, fix equipment, deal with sickness or illness, grow their own food and recycle their waste. They might use Martian soil to create rocket fuel for their return, or turn it into useful things by 3D printing.

When Earth and Mars are on opposite sides of the Sun, the distance from one to the other is about 400 million kilometres, and it takes radio signals twenty-four minutes to go between them. This means that astronauts on Mars won't be able to speak directly to anyone on Earth, having to wait a long time for answers to come back from mission control. They would only be able to communicate with their families via emails or recorded messages.

Engineers and scientists need to figure out all these things before humans can walk on Mars. The crew of perhaps six people will be supported by a huge team worldwide, doing every sort of job imaginable, all dedicated to exploring Mars, carrying out scientific experiments, and bringing the crew back safely. They will need to be determined, patient and passionate about space exploration.

You can be a part of that team. With ambition, hard work and determination you can do anything you set your mind to. So what are you waiting for? The Red Planet is waiting for you...

GALLERY OF ILLUSTRATORS

THE
ORIGINS
OF
**SPACE
TRAVEL**
→

ÉMILIE du CHÂTELET
18
ADI PASHALIDI
KOZELJ

ADA LOVELACE
20
AMY McCARTHY

JEANNETTE PICCARD
22
ANA RÍOS WANG

MARY SHERMAN MORGAN
24
CÉLESTE MUETH

JACQUELINE COCHRAN
26
LAUREN HACKETT

THE
DAWN
OF THE
**SPACE
AGE**
→

VALENTINA TERESHKOVA
32
EHIGIE AIGIOMAWU

JERRIE COBB
34
ELEANOR CREWES

THE MERCURY 7 WIVES
36
YENI OGUNMILADE

EILENE GALLOWAY
38
SARAH HALLAM

MARY JACKSON
40
ROSIE CHOMET

DEE O'HARA
42
JACK MERTON

KATHERINE JOHNSON
44
LIAM JAMES SPROSTON

MARGARET HAMILTON
46
EMILIA BOCIANOWSKA

THE WALTHAM 'LITTLE OLD LADIES'
48
JOHANNA BLAHA

POPPY NORTHCUTT
50
HUONG ANH TRINH

RITA RAPP
52
CLARISSE HASSAN

DOTTIE LEE
54
AMY McCARTHY

THE ILC SEAMSTRESSES
56
SIYAKORN KIMHASAWAD

SPACE STATIONS AND SHUTTLES
→

SALLY RIDE
62
JULIETTE STUART

SVETLANA SAVITSKAYA
64
MIRANDA SMART

NICHELLE NICHOLS
66
EHIGIE AIGIOMAWU

CHRISTA McAULIFFE JUDY RESNIK
68
LIAM JAMES SPROSTON

MAE JEMISON
70
HUONG ANH TRINH

HELEN SHARMAN
72
SIYAKORN KIMHASAWAD

EILEEN COLLINS
74
SHANA PAGANO-LOHREY

CHIAKI MUKAI

76

HUONG
ANH TRINH

CLAUDIE HAIGNERÉ

78

JOHANNA
BLAHA

PATRICIA COWINGS

80

ANA
RÍOS WANG

IRENE LONG

82

ELEANOR
CREWES

LIVING AND WORKING IN SPACE
→

PEGGY WHITSON

88

ROSIE
CHOMET

JULIE ROBINSON

90

WIKTORIA
CWIEK

SUNI WILLIAMS

92

NICOLA
REEVES

JEANNE LEE CREWS

94

AMY
McCARTHY

KALPANA CHAWLA

LAUREL CLARK

96

JULIETTE
STUART

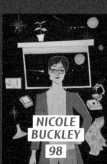

NICOLE BUCKLEY

98

ADI PASHALIDI
KOZELJ

BERTI BRIGITTE MEISINGER

100

JOHANNA
BLAHA

SAMANTHA CRISTOFORETTI

102

SARAH
HALLAM

NORIKO SHIRAISHI

104

LIAM JAMES
SPROSTON

ANOUSHEH ANSARI

106

CLARISSE
HASSAN

GINGER KERRICK

108

SIYAKORN
KIMHASAWAD

THE
FUTURE
OF
SPACE
→

ELENA SEROVA
110

KATYA
KOROLEVA

LIU YANG
WANG YAPING
116

WIKTORIA
CWIEK

KELLY
LATIMER
118

SUN YOUNG
KIM

GWYNNE
SHOTWELL
120

YANG
YANG

FRANCES
WESTALL
122

ELEANOR
CREWES

SIMONETTA
di PIPPO
124

ZOE
LANDWEHR

ELLEN
STOFAN
126

EMILIA
BOCIANOWSKA

MONICA
GRADY
128

ROSIE
CHOMET

GERDA
HORNECK
130

JACK
MERTON

ANITA
SENGUPTA
132

CÉLESTE
MUETH

THE FIRST
PERSON
ON MARS
134

KATERINA
DEMETRIOU-JONES

YOUR MISSION

So now you know that there's nothing standing in your way if you decide you want to help humankind take its next steps in exploring the universe, or even go into space yourself one day. If you think that's going to be your mission in life, you can start by filling out this plan:

>YOUR NAME

>FLAG   >NATIONALITY

D D M M Y Y Y Y

>DATE OF BIRTH

>WHO MOST INSPIRES YOU IN THIS BOOK

The people you have read about have all sorts of interests and careers. Tick the boxes below of the jobs that you think might help you on your mission into space:

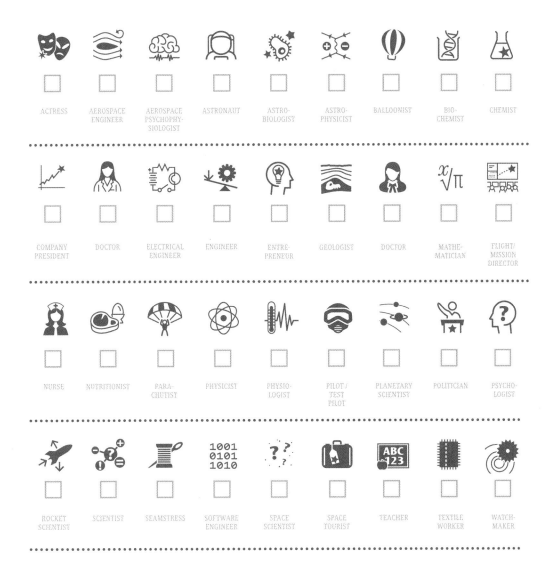

ACTRESS · AEROSPACE ENGINEER · AEROSPACE PSYCHOPHYSIOLOGIST · ASTRONAUT · ASTRO-BIOLOGIST · ASTRO-PHYSICIST · BALLOONIST · BIO-CHEMIST · CHEMIST

COMPANY PRESIDENT · DOCTOR · ELECTRICAL ENGINEER · ENGINEER · ENTREPRENEUR · GEOLOGIST · DOCTOR · MATHE-MATICIAN · FLIGHT/MISSION DIRECTOR

NURSE · NUTRITIONIST · PARACHUTIST · PHYSICIST · PHYSIOLOGIST · PILOT / TEST PILOT · PLANETARY SCIENTIST · POLITICIAN · PSYCHO-LOGIST

ROCKET SCIENTIST · SCIENTIST · SEAMSTRESS · SOFTWARE ENGINEER · SPACE SCIENTIST · SPACE TOURIST · TEACHER · TEXTILE WORKER · WATCHMAKER

ABOUT THE AUTHOR

LIBBY JACKSON is one of Britain's foremost human space flight experts, and is the Human Space Flight and Microgravity Programme Manager for the UK Space Agency. Libby's career working in the space industry began when she applied for work experience at NASA aged seventeen from her secondary school in Kent. Weeks later she was sitting in mission control in Houston. Ten years on, and after completing a physics degree at Imperial College, she was back working at mission control for the European Space Agency side in Munich. She was an instructor, a flight controller and finally a Columbus flight director on missions to the International Space Station. From 2014 to 2016, she managed the hugely successful UK Space Agency education and outreach programme that supported Tim Peake's mission.

@LibbyJackson__   LibbyJackson

ABOUT THE ILLUSTRATORS

The illustrators are students and graduates from BA (Hons) Illustration and Visual Media at the **LONDON COLLEGE OF COMMUNICATION (LCC)**, part of the University of the Arts London (UAL), which is a pioneering world leader in creative communications education, preparing students for successful creative careers. LCC courses are known for being industry focused with students taught by an inspiring community of experienced academics, technical experts and leading specialist practitioners. Generations of award-winning photographers, filmmakers, screenwriters, journalists, broadcasters, designers and advertising and PR professionals have started their careers at LCC, and today's graduates continue to be highly sought after and win prestigious international awards.

www.arts.ac.uk/lcc

ACKNOWLEDGEMENTS

This book would not have existed were it not for the exceptional levels of support and belief of those at Penguin Random House and the UK Space Agency; Ben Brusey, Lizzy Gaisford, Katie Loughnane, Matt Goodman, Sue Horne and Diana Martin.

Berti Brigitte Meisinger, Nicole Buckley, Anita Sengupta and Kelly Latimer very graciously and generously shared their time with me; thank you very much, it was an honour to hear you share your stories.

I am indebted to all the brilliant people who offered information and suggestions in my research, particularly George Morgan, Glen Cornhill, Dallas Campbell, Kevin Fong, Gale Allen and Elizabeth Scully. Your suggestions and help have all played their part in shining the spotlight on these amazing women.

My friends and family have been magnificent, always offering encouragement and understanding through the project, even when I seemed to have vanished into hibernation.

And to Chris, for everything.

CENTURY

20 Vauxhall Bridge Road
London SW1V 2SA

Century is part of the Penguin Random House group of
companies whose addresses can be found at
global.penguinrandomhouse.com.

 Penguin
Random House
UK

Copyright © Libby Jackson 2017

Libby Jackson has asserted her right under the
Copyright, Designs and Patents Act, 1988,
to be identified as the author of this work.

Illustrations © individual illustrators 2017
as detailed on pages 136–139

First published in 2017 by Century

www.penguin.co.uk

A CIP catalogue record for this book is available
from the British Library.

ISBN 9781780898360

Design and chapter opener illustrations
by Tim Barnes ⌖ herechickychicky.com

Typeset in Generis Slab 10/16pt, Mosquito and Eagle
by Monotype

Printed and bound in Italy by L.E.G.O. S.p.A.

Penguin Random House is committed to a sustainable
future for our business, our readers and our planet.
This book is made from Forest Stewardship Council®
certified paper.